Wakefield Press

Wine, Terroir and Climate Change

John Gladstones is a leading Australian agricultural scientist, with a distinguished record in the breeding, agronomy and botany of crop and pasture legumes that has earned him many scientific and community awards, including Member of the Order of Australia (AM). His pioneering work in viticulture led to the establishment of Margaret River as a premium wine-producing region. His earlier book *Viticulture and Environment* (1992) was awarded 'Special Distinction in Viticulture' by the Office International de la Vigne et du Vin, Paris. He lives in Perth with his family.

Wine, Terroir and Climate Change

John Gladstones

Wakefield Press

Wakefield Press
1 The Parade West
Kent Town
South Australia 5067
www.wakefieldpress.com.au

First published 2011

Cover design by Stacey Zass, Page 12
Cover photograph by Peter Dry
Typeset by Wakefield Press
Printed and bound by Hyde Park Press, Adelaide

National Library of Australia Cataloguing-in-Publication entry

Author: Gladstones, J.S. (John Sylvester), 1932–   .
Title: Wine, terroir and climate change / John Gladstones.
ISBN: 978 1 86254 924 1 (pbk.).
Notes: Includes bibliographical references.
Subjects: Viticulture – Climatic factors.
                  Vineyards – Location.
                  Viticulture – Environmental aspects.
Dewey Number: 634.8

Government
of South Australia

Arts SA

fox creek
wines

Contents

Preface

The stimulus for this book came out of an invitation to present a paper at the Viticultural Terroir Conference held at the Davis campus of the University of California in March 2006. In the event I was unhappy with the coverage I was able to give to my topic, and noted that although many of the other papers were interesting and worthwhile, most dealt only with limited aspects. Nor did the extant literature contain much on terroir that was enlightened by modern research. Older French writings were still valuable: indeed, I found little in them to contradict. But questions remained as to details and explanation, and to how universally the French experience applied. Moreover, the concept of terroir was becoming overshadowed by fears of drastic climate change.

All this posed an irresistible challenge. My earlier book *Viticulture and Environment* (1992) had covered the subject in part, but much had happened since. I therefore resolved to explore more deeply the topics of both terroir and climate change, and some interrelations between them promised to throw new light. In doing so I have tried to follow only the scientific and historical evidence, and what flows logically from it. This book is the result.

The new research for it was conducted during tenure of an Honorary Research Fellowship at the University of Western Australia, Perth. I particularly acknowledge the facilities and staff help of its Biological Sciences, Physical Sciences and Reid libraries, and of the library of Curtin University, Perth, which among them carried nearly all the relevant scientific journals.

My thanks are due to Caroline Wallace (née Criddle), June Thom-Allen and Val Hall for secretarial assistance at various stages, and to Neil Delroy and Dennis Criddle who in part facilitated it; also to Steve Barwick for his work on the illustrations. My daughter Helen helped with some of the proof-reading. Peter Dry kindly provided the front cover illustration. And it was a pleasure again to work with Michael Deves as editor, who saw the book through press in his usual professional manner and mostly tolerated my idiosyncratic style preferences.

Above all I thank my wife Pat for her unfailing support and encouragement, both through the research and writing of the book, and through a long and often distracting scientific career. To her I gratefully dedicate this volume.

J.S. Gladstones
Perth, December 2010

Chapter 1

Introduction and Definition of Terroir

This book tackles two contentious subjects that underlie the future of viticulture. Terroir is much spoken of, but nobody, to the best of my knowledge, has attempted a comprehensive definition and integration of its elements in the light of modern science. To do so is an ambitious task, given the many remaining gaps in knowledge. Some of my conclusions may prove to be wrong. But I trust at least that they will help lead to a fuller understanding.

Climate change, which takes up much of the book's latter half, must obviously influence all planning for future viticulture. But in approaching the subject it became evident that neither public understanding nor the 'official' position of the Intergovernmental Panel on Climate Change (IPCC) was necessarily accurate. Much in the argument for global warming by anthropogenic (man-caused) greenhouse gases appeared questionable. I therefore undertook as deep a study of the basic scientific evidence as I was able. The result was disturbing, though more as to the science underlying the global warming thesis than to the future of viticulture.

Establishing my conclusions on climate unavoidably requires extensive referencing. Some readers will prefer to by-pass this, but I would encourage those seriously interested to look further for themselves. Most is in standard scientific books and journals, and readily enough accessible. Some of the most interesting evidence comes from viticultural history.

The French term *terroir* has no exact equivalent in English, and when transferred to the English language has been given a bewildering array of meanings depending on user perspective. Turner and Creasy (2003) discuss these in detail. In *Viticulture and Environment* (1992) I deliberately avoided the word for that reason, although the book was in fact largely about terroir.

Since then the word and the concepts it conveys have become mainstream world-wide.

Here I use the term in what I believe is its original and correct sense, as set out by French writers such as Laville (1990). That is, simply, the vine's whole natural environment, the combination of climate, topography, geology and soil that bear on its growth and the characteristics of its grapes and wines. Local yeasts and other microflora may also play a part. Indeed, the naturally-occurring yeasts on grapevines and around the wineries may well prove to play an important, and hitherto unappreciated, role in the subtler aspects of wine terroir characteristics both locally and regionally. I do not discuss them further here, but future research and practical experience could add much of interest, particularly for the New World where the use of cultured yeasts has become the norm. All these factors interact with management in the vineyard and winery to shape the wine. Treatment (or mistreatment) in storage and commerce then further influences final drinking qualities. But these latter influences are not properly part of terroir. To include them creates complexities that largely preclude terroir definition.

Two part-exceptions must be admitted. The first is soil modification by, for instance, drainage (as in Bordeaux), terracing, or progressive fertility change related to soil management. But these in turn become semi-permanent features of individual sites, and can broadly be considered to become parts of their terroirs. The second is possible man-caused climate change.

Terroir, then, describes the unique geography of a wine's origin. It is not a property of the wine itself. Good wine *reflects* the terroir(s) of its origin.

Terroir scale varies depending on its controlling factors. A defined terroir can range across many kilometres if the land is flat and there is little variation in climate, geology and soil. More broadly it can encompass an entire region with substantially uniform soils and climate. On the other hand it can be confined to within tens or hundreds of metres, as in Burgundy, where localized soil and drainage differences can be decisive. In practice there must be some flexibility of definition, depending on site variability and commercial purpose.

The important thing is that a wine's defined origin conveys a meaningful message to buyers and consumers, mostly as to its style though not necessarily as to quality, which depends on other factors as well. (The more restrictive European appellation schemes try to combine the two concepts, but with limited success.) Obviously a detailed assurance of a wine's origin is critical for wines of great repute, individuality and price. But it remains important also for lesser wines defined simply by grape variety (or blend)

and region, which will comprise much of the commercial wine of the future. Some predictability of character is still needed for these if they are to achieve market differentiation, recognition and success.

🙣

Several recent publications have concentrated on particular aspects of terroir, e.g. Pomerol (1989) and Wilson (1998) on geology, White (2003) on soils, and my own writings (Gladstones 1992, 2004) principally on climate. All have been useful, but none has fully encompassed the complex interactions that go to make up terroir. That is what I attempt in this book.

The book's plan is first to deal with climate in its broad sense, starting with the central role of temperature (Chapter 2), then its other elements (Chapter 3). Chapter 4 looks at geographic effects on macroclimate or regional climate, followed by those of local factors on mesoclimate. The coverage to there is on similar ground to that of my previous publications (Gladstones 1992, 2004), but with more specific focus on terroir. It also brings into account some important later research.

The following several chapters delve more deeply into terroir as an integrated concept taking in the vine, climate, the soil and its underlying geology. Central to this discussion is the development of a hypothesis that relates grape ripening to root-produced hormones, influenced by a combination of both soil and atmospheric conditions. Chapter 9 brings in organic and biodynamic viticulture.

Chapter 10 presents a revised and expanded list of grape maturity groups, a necessary provision for predicting maturity dates and ripening conditions for varieties across the range of climates. It also touches on some implications of maturity rankings for wine style.

Chapter 11 describes a revised method for constructing comprehensive viticultural climate tables, including estimated average maturity dates and ripening conditions for the respective grape maturity groups. As compared with that previously described in *Viticulture and Environment* (1992) the method is (I hope) a little clearer and more logical. It introduces the additional criterion of cloudiness, which recent evidence has suggested more and more to be an important terroir descriptor. There are also new indices for spring frost risk and summer heat stressfulness. Appendix 2 gives reference examples of completed tables, while Table 3.1 lists suggested 'ideal' sets of ripening conditions for the different wine styles, against which any site's estimated ripening conditions for each grape maturity group can be compared.

Chapters 12 and 13 deal with the prospects of climate change and its

What defines terroir influences on say Prima of mandwic, Colli del Groic?. also a als zones Salento, Bari, Bren?

potential impact on viticulture. Finally Chapter 14 paints a speculative picture of viticulture's global future in the 21st century.

One point needs to be made for the sake of clarity. Throughout I use the term 'mean' in its strict sense, i.e. as being half way between two (and only two) extremes. Thus a day's mean temperature is its (maximum + minimum)/2. An average can be that of any number of values. A day's true average temperature can only be derived from continuous recording, or, less accurately, from recordings at twenty-minute, hourly or other intervals; the results differ significantly among themselves and from the mean. While true averages may be more accurate for detailed local or within-season studies, I have preferred to use means for the practical reason that only records of maxima and minima are as yet widely enough available, from long enough records, to give comprehensive and reliable world-wide comparisons. Monthly, seasonal or annual average means, then, are here the averages of daily means over the specified periods.

Above ~40°C photsyn shuts down

what is optimal → R ↑ plant?

what does curve look like?　　max days (hrs)

effect longer term on °C ↑ shut off?　　at optimal

Temp - varietal of grape maturation?

Chapter 2

Temperature: The Driving Force

2.1   Temperature and vine phenology

Temperature is central to all aspects of viticulture. The evidence is now clear that, with only minor other influences, it alone controls vine phenology, i.e. the vine's rate of physiological development through budbreak to flowering, setting, veraison, and finally fruit ripeness. Light interacts with temperature to govern photosynthesis, dry matter production and potential yield; but as will be discussed in subsection 2.1.3, it does not bear directly on phenology.

High and low extremes of crop load and water availability can advance or retard veraison a little, but within environments and with management for quality wine production these differences are mostly very small. In that context they can safely enough be neglected for the purpose of predicting average phenology from average temperatures.

The relationship between phenology and temperature is not linear. But with certain adjustments to recorded temperatures, based partly on known plant physiology and partly on practical observation, it is possible to estimate 'biologically effective' temperatures and heat summations that do give a linear fit across more or less the full range of viticultural environments. This section describes these adjustments.

2.1.1   *The 19°C mean temperature cap*

As will be discussed in Section 2.2, growth of vines and most other temperate plants *as measured by dry matter increase* rises from nil a little below 10°C mean temperature, reaches a maximum at means around 22–25°C, and falls again to nil as means reach about 40°C (Figure 2.1).

However, rate of phenological development, measured by production rate of new stem nodes and times between phenological stages, follows a different response pattern. It is unrelated to photosynthesis and dry matter production.

5

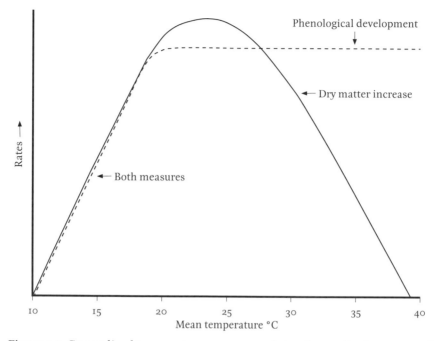

Figure 2.1. Generalized temperature responses of vine dry matter increase and phenological development.

As Figure 2.1 shows, it responds similarly to mean temperatures up to about 20°C, then plateaus. Buttrose (1969), Buttrose and Hale (1973) and Schultz (1993) all present evidence showing such a plateau response for vines.

The shape of the phenology response curve can be represented, as a first approximation, by a positive straight-line response in the lower temperature range and flat above a mid-temperature inflection point (Figure 2.1). While this does not fit the true curve perfectly, it does so with minimal error when averaged over the seasonal range of mean temperatures that grapevines normally encounter. Temperatures effective for predicting maturity dates can be fairly approximated by simple capping at a temperature giving best overall fit to the curve.

In developing the concept for *Viticulture and Environment* (1992) I tried matching different inflection temperatures against known combinations of climate and vine phenology throughout the world, measured by average maturity dates of known grape varieties for dry table wines. Inflection at 19°C mean temperature gave the best match, and proved to be a serviceable starting point for estimating grape maturity dates over the range of climates reported in that work.

Two corollaries are in order. First, the shape of the phenology response curve explains why neither raw (uncapped) degree days, such as those of Amerine and Winkler (1944), nor the curve of dry matter increase, is useful for predicting vine phenology. As a result many viticultural researchers, e.g. McIntyre et al. (1987), have tended to dismiss the relevance of temperature summations to vine phenology. We can now see this to have been mistaken.

The second corollary is that a 19°C mean temperature cap goes far to explaining a widely observed phenomenon: that temperatures of the first two or three growing season months, or alternatively the date of flowering, can usually predict quite closely the dates of veraison and maturity to follow. This comes from the fact that in most climates the temperatures up to flowering are in the range to which phenology is highly responsive. After that they are mostly in, or close to, the range of flat response. The later phenological intervals therefore show little response to temperature, and tend to be constant from year to year.

2.1.2 Adjustment for diurnal temperature range

Discussion so far has been in terms simply of mean temperature, i.e. (maximum + minimum)/2. Many studies, mostly in relation to greenhouse floriculture, have shown that where diurnal temperature ranges are narrow, as is normal in such culture, mean temperatures quite accurately predict rates of node or leaf appearance. The relationship continues to apply where controlled night temperatures are higher than those during the day, as is now often done to produce compact potted plants with short internodes: see Karlsson et al. (1989); Grimstad and Frimanslund (1993); Myster and Moe (1995). Besides confirming the primacy of temperature in controlling phenological growth processes, their data show that these proceed continuously day and night and that day and night temperatures control in identical ways.

Plants growing in the open generally experience wider and much more variable temperature ranges than those in greenhouse culture, with more likelihood of effects on plant development. The early phytotron research of Went (Went 1953, 1957; Went and Sheps 1969) showed that, with some variation among plant species, a narrow diurnal range is optimal for growth and that night temperatures are critical. Wide ranges retard development.

The shape of the grapevine phenological response curve to mean temperatures (Figure 2.1) can readily be related to these findings. A 24-hour period with wide diurnal variation can experience day temperatures well into the plateau response range, whereas plunging night temperatures reach into the range of greatest restriction by temperature. The latter must then be the

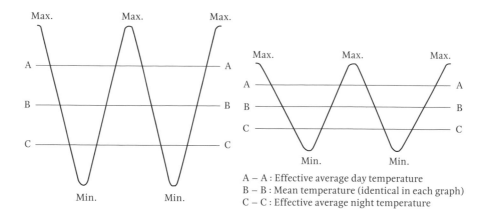

A – A : Effective average day temperature
B – B : Mean temperature (identical in each graph)
C – C : Effective average night temperature

Figure 2.2. Effect of diurnal temperature range on effective night and day temperatures. From Gladstones (1992), after Went (1957).

prime limiting factor.

Langridge and McWilliam (1967) reported that control by night and/ or minimum temperatures is indeed common among temperate plants. It makes particular evolutionary sense for perennial deciduous species that recommence growth in spring, by helping to delay budbreak until the worst danger of frosts is over. After budbreak, a slowing of growth by low night temperatures favours the accumulation of protective compounds in the new tissues that will enhance their resistance to later frosts. Field experience suggests that grapevines conform to this pattern.

Figure 2.2, after Went (1957), shows the relationship between diurnal range and effective night temperature, which is half way between the minimum and the mean. For every 1°C increase in diurnal range, effective night temperature falls by 0.25°C.

In *Viticulture and Environment* (1992) I used this to adjust for diurnal range in calculating monthly effective temperature summations; but because its application across all diurnal ranges appeared to over-compensate, I confined it to the widest and narrowest ranges. That is, for every 1°C wider range than 13°C the effective mean reduced by 0.25°C, while for every 1°C narrower range than 10°C it increased by 0.25°C. This procedure significantly improved the fit between climate data and observed average ripening dates. Capping the resulting effective means at 19°C automatically confined the influence of diurnal range to months when temperature directly limits phenological development.

In the present work I have modified the adjustment to be simpler and seemingly more logical. Comparable adjustments apply over all diurnal ranges, subtracting from the mean 0.25°C per 1°C wider range than 12°C and adding 0.25°C per 1°C narrower range than 12°C; but they do so only for the first four months of the growing season, i.e. April–July in the Northern Hemisphere and October–January in the Southern Hemisphere. There are two reasons for this.

First, these months include the periods of budbreak and early spring growth, when the retarding effect of low minima is ecologically most logical and best established. They extend to about the latest dates (in cool climates) when growth of the fruit is by cell division: a process that continues through the night and may tend to be concentrated then. Capping the resulting monthly effective means at 19°C again ensures that the adjustment registers only in months with low enough mean temperatures to retard phenological development. In warm to hot climates this will include no more than the first month or two of the growing season.

Second, no such clear argument exists for the ripening period. It is true that the metabolism of forming flavour compounds, which we can assume to continue day and night, is likely to be most limited by low night temperatures. Also, high day temperatures can be counter-productive through evaporative or degradative losses of flavour components and pigments. We will examine these aspects more closely in Section 2.3. But contrary to flavour ripening, sugar ripening depends primarily on daytime warmth and sunshine. For the total process of ripening, therefore, neither day nor night temperatures can be claimed as definitive.

There is furthermore the special case of ripening at viticulture's cool limit, where often much of the night is too cold for ripening activity of any kind. Ripening then depends more or less entirely on daytime warmth and fruit sunshine exposure, both of which are best provided by sunny weather. This, for a given site and time of season, tends to have the widest diurnal temperature range.

Given these mixed responses, phenology post mid-summer seems best related simply to the temperature means. Combined with adjustments as just described for the growing season's first four months, and for daylength as described below, this appears to have given at least as good and probably a better fit between climate and average grape maturity dates than my previous method.

2.1.3   Adjustment for daylength

There is universal agreement in the European literature that the rate of

vine phenological development varies in proportion to a product of temperature and daylength. The Heliothermic Index of Branas (1946) multiplies
summations of mean temperature over a 10°C base by daylengths through
the season, and has long been accepted as defining the northern limit of
viticulture in Europe. That of Huglin (1983) uses instead summations based
on effective daytime temperatures, i.e. half way between the means and
the maxima. Its argued rationale is that phenology and dry matter growth
depend alike on photosynthesis, and therefore that the operative temperatures are those of the day. Also, because photosynthesis reaches a maximum
at light intensities well below those of full sunlight, daylength is more important than hours of bright sunshine.

But as we saw in subsection 2.1.2 above, and as found in much field crop
research not cited here, rates of plant phenological development as measured by those of node or leaf appearance depend quite strictly on temperature, with no significant influence of either light intensity or its duration.
(Photoperiod does govern times of flower initiation in many plant species,
but that is not relevant to vine phenology.) In the grapevine, experimental
results such as those of Buttrose (1969) and Schultz (1993) confirm the lack
of any relationship of phenology at least to light intensity. To the extent of
such a seeming relationship in northern Europe, part is probably because
maturation there is as often limited by slow sugar accumulation as it is by
true 'physiological', or 'flavour', ripening. Also there are reasons to suggest
that much of the apparent association with daylength results from indirect
relationships to temperature. At least two mechanisms exist for this.

The first is that adjustments proportional to daylength correct an inac-
curacy caused by using temperature means rather than true averages. Under
long days, temperatures will tend to plateau close to the maximum for longer
than in short days. The mean, or (maximum + minimum)/2, then under-
estimates true daily average temperature as would be derived by continuous
measurement. The opposite happens under short days.

The second is that phenology will logically be more directly related to
vine and fruit temperatures than to air temperatures. This applies especially
to fruit ripening. Adams et al. (2001) showed that expansion and ripening
of tomato fruits depended on the temperatures of the fruits, not those of
the air or leaves. A similar relationship seems likely for grapes, although
the literature has little to say on the point. Certainly, as Smart and Sinclair
(1976) have shown, sun-exposed grape clusters attain temperatures many
degrees above those of the air or the leaves, because unlike the latter, the
berries have few stomata for cooling transpiration. Moreover, berry and

Table 2.1. Multipliers for adjusting effective heat (°C minus 10) for daylength, based on month and latitude[1].

Month/ Lat.	N. April S. October	May November	June December	July January	August February	September March	October April	Average
25°	0.962	0.928	0.912	0.920	0.946	0.987	1.038	0.956
30°	0.972	0.951	0.938	0.942	0.961	0.991	1.027	0.969
35°	0.986	0.974	0.967	0.969	0.979	0.995	1.014	0.983
40°	1.000	1.000	1.000	1.000	1.000	1.000	1.000	1.000
42°	1.007	1.013	1.015	1.014	1.009	1.002	0.995	1.008
44°	1.014	1.026	1.031	1.028	1.019	1.005	0.989	1.016
46°	1.022	1.040	1.049	1.045	1.029	1.008	0.982	1.025
48°	1.030	1.055	1.069	1.062	1.041	1.011	0.974	1.035
50°	1.039	1.072	1.090	1.082	1.053	1.014	0.966	1.045
52°	1.049	1.090	1.113	1.103	1.066	1.018	0.957	1.057

[1] Calculated from the *Smithsonian Meteorological Tables*, Sixth Revised Edition, 1951. The Smithsonian Institution, Washington.

cluster geometry means that their heating is cumulative, so that the duration of daytime heating contributes to the temperatures they reach and how long these are maintained, and thus, very probably, to effective temperatures for fruit development.

Whatever the mechanism, I continue to follow the procedure of Branas (1946) of multiplying monthly mean temperature averages in °C minus 10 ('effective heat') by factors proportional to the respective monthly average daylengths. Here, as in *Viticulture and Environment* (1992), I take latitude 40° as a neutral pivotal point, multiplying (from spring to autumn equinox) by factors greater than 1 at higher latitudes and less than 1 at lower latitudes. Table 2.1 sets out factors for selected latitudes. Graphing these provides curves for convenient reading of values for intermediate latitudes.

Simplified, the total procedure of estimating biologically effective degree days is to take monthly average mean temperatures, adjust these for diurnal range as in subsection 2.1.2, then multiply the resulting mean temperature excesses over 10°C by the respective monthly daylength factors for the site latitude. Each month is subject to final capping at 19°C. Chapter 11 treats the steps in full detail and gives an example of their working. From here on I shall refer to the resulting biologically effective degree days as E°days.

2.1.4 Growing season temperature summations
Definitions of potential length of the vine growing season have varied. The commonest, which I adopt here, is the seven months April–October inclusive

in the Northern Hemisphere, or October–April in the Southern Hemisphere. Note that all present calculations and adjustments assume that definition. Used with other assumed growing season lengths they will not give as accurate or universal a fit to vine phenology.

The primary use of E°days is to estimate average dates of fruit maturity in any climate: maturity being defined here as readiness to pick for making dry table wines. Table 10.1 gives a revised and expanded classification of *vinifera* grape varieties into nine maturity groups, from earliest (Group 1) to latest (Group 9), together with the numbers of E°days from 1 April/1 October to bring each to maturity. Estimating the maturity dates requires cumulative accounting of E°days through the months and finally the days of the month in which maturity occurs. Section 11.7 describes a graphic method for this.

A problem arises in estimating seasonal temperature summations in cool climates where the curve of monthly average mean temperatures passes through 10°C within April and/or October. Were it to do this mid-month, giving an average mean of 10°C for the month, no °days would accrue; yet half the month is above 10°C and doubtless contributing to growth or maturation. Calculation of °days in this case is as shown in Figure 2.3: either from the total shaded area under the curve or, accurately enough, by taking the centre point of the period within the month above 10°C, reading off its temperature and multiplying by the number of days above 10°C.

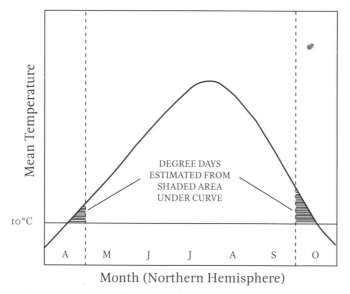

Figure 2.3. Estimation of April and October temperature summations from the temperature curve in cool climates. From Gladstones (1992).

2.1.5 *Adjustments for vineyard sites and climate change*

Some weather recording stations with reliable long-term records represent well enough the conditions of vineyards in their vicinities. Their averages can be used directly to estimate grape maturity dates and, from them, average conditions for the ripening period. As will be detailed in Section 2.3, for comparative purposes I take this to be the last 30 days to the estimated date of maturity.

In fact few vineyards are so situated. Most are at a distance from weather stations with reliable long-term records, and most are at different altitudes. But both these factors can be reasonably allowed for: the first by interpolation between or among multiple stations, or from climate maps, as described in Chapter 11; the second, by accounting altitude differences at 0.6°C per 100 metres. Both corrections apply alike to the monthly average means and to the maxima and minima. Short-term data from a vineyard or near-vineyard site, say from 3–4 years or more, may also be corrected by calibration against simultaneous records from a related but more distant site with good long-term records. In estimating E°days the resulting climate estimates receive adjustments for diurnal range, daylength and 19°C mean temperature cut-off in the same way as for directly recorded data.

These calculations serve well enough in low-latitude country with flat terrain and uniform soils. But Old World vineyards, especially at high latitudes, are mostly in hilly country and depend very much on individual site characteristics for their wine qualities and often for their ability to ripen grapes at all. Increasingly the same can be said of the New World, as it seeks out its best cool-climate terroirs for quality wines. Generalized estimates cannot describe adequately the climates of these specialized sites. To do that requires that their individual terroirs be accounted for.

Base weather stations at the highest viticultural latitudes already commonly gain around 100 E°days compared with the raw temperature summations through adjustments for their latitude and diurnal range as already described: see, for instance, the table for Geisenheim, Germany, in Appendix 2. Necessary vineyard site selection in the most cold-marginal climates may then add a further 100–200 E°days through allowing for altitude differences, and for other topographical factors such as slope and aspect that will be dealt with in Chapter 4 in detail.

Even in the absence of topographical differences, well-drained stony or rocky soils are observed universally to advance grape maturity compared with cold, water-retentive loamy soils. The results of Tesic et al. (2001a, 2001b) in New Zealand provide a good example of this. Such observations make it possible to estimate equivalent temperature adjustments for all major site factors as best they can be judged: whether for individual sites or for typical vineyard sites within an area. Ultimate testing comes from the goodness of fit between the resulting estimates and observed maturity dates across the full range of viticultural regions having well documented terroirs (e.g. Johnson and Robinson 2001). That process underlies the site adjustments as used in *Viticulture and Environment* (Gladstones 1992) and further developed here.

Table 2.2 sets out in summary the revised adjustments for individual site factors. These are guidelines only: intermediate values can be used as appropriate. Full values imply strong expression of a factor within its normal range.

All values are simply additive, giving a net sum to be added to a site's raw monthly temperature averages, treating maxima and minima separately.

The revised adjustments for slope, aspect and soil apply only to the minimum temperatures, recognizing the fact that they reflect primarily the drainage of cold air at night or the night re-radiation of soil-absorbed heat. There is little evidence that any of these factors greatly influence maximum temperatures, which require only the universal altitude adjustment of 0.6°C per 100 metres and in any case probably have minimal effect on phenology.

Adjustments for proximity to/distance from water bodies (see Section 4.6) must take into account the relative extents of land and water and the surrounding topography that limits their local air convections. Although temperature means remain unchanged, the higher minima and reduced diurnal range close to water do significantly raise E°days and advance phenology, as is commonly observed. They also have an important influence in reducing temperature extremes and the various indices of temperature variability to be described in subsection 2.4.2.

The new phenological factor of exposure to/isolation from sea breezes and regionally prevailing cool winds, which is discussed in Section 3.4 more fully, applies mainly to exposed coastal regions that are backed by warmer and drier hinterlands. A fairly clear case of effective cooling beyond that indicated by the temperature means is along the south coasts of mainland Australia and South Africa, where on-shore summer trade winds result in the early onset of afternoon sea breezes and much reduced temperatures for the rest of the day. The opposite effect of isolation from cool winds (of any kind) is seen typically on the protected, sun-exposed sides of hill or mountain ranges, where up-slope afternoon-evening convections can prolong exposure to winds coming from warmer inland or lower-latitude climates. Section 3.4 deals gives possible examples.

Table 2.2. Guidelines to temperature adjustments, °C, for site factors other than altitude[1].

Adjustment factor	Min.	Max.	Mean	Diurnal Range
Slope and air drainage				
Strong cold air ponding	−1.6		−0.8	+1.6
Moderate cold air ponding	−0.8		−0.4	+0.8
Flat or free-draining valley	−	−	−	−
Moderate slope	+0.4		+0.2	−0.4
Moderate slope on isolated hill	+0.8		+0.4	−0.8
Steep slope	+0.8		+0.4	-0.8
Steep slope on isolated hill	+1.2		+0.6	−1.2
Inclination to midday sun[2]				
On moderate slope, <40° latitude	+0.2		+0.1	−0.2
On moderate slope, 40 - 48° latitude	+0.4		+0.2	−0.4
On moderate slope, >48° latitude	+0.6		+0.3	−0.6
On steep slope, <40° latitude	+0.4		+0.2	−0.4
On steep slope, 40 - 48° latitude	+0.8		+0.4	−0.8
On steep slope, >48° latitude	+1.2		+0.6	−1.2
Soil				
Deep, loamy, water-retentive	−0.4		−0.2	+0.4
Mod. stony or calcareous, free draining	+0.4		+0.2	−0.4
Very stony or calcareous, very free draining	+0.8		+0.4	−0.8
Proximity to/ distance from water bodies				
Moderately closer to lake etc.	+0.4	−0.4		−0.8
Much closer to lake etc.	+0.8	−0.8		−1.6
Moderately closer to ocean	+0.8	−0.8		−1.6
Much closer to ocean	+1.2	−1.2		−2.4
Moderately further from lake etc.	−0.4	+0.4		+0.8
Much further from lake etc.	−0.8	+0.8		+1.6
Moderately further from ocean	−0.8	+0.8		+1.6
Much further from ocean	−1.2	+1.2		+2.4
Exposure to/ isolation from prevailing cool winds[3]				
Moderate exposure		−0.8	−0.4	−0.8
Strong exposure		−1.6	−0.8	−1.6
Moderate isolation		+0.4	+0.2	+0.4
Strong isolation		+0.8	+0.4	+0.8

[1] Allow also 0.6°C per 100 metres altitude difference to both minimum and maximum. All adjustments fully additive. Mixed or intermediate ratings *pro rata*.

[2] Full allowance for directly or near-directly facing the midday sun. Opposite aspects negative, E and W aspects neutral.

[3] Applies mainly to marine wind exposure and inland isolation behind hill or mountain ranges. See Section 3.4 for full discussion.

Finally, how can prospective climate change be accommodated? That is simply done. For any site it is necessary only to add whatever temperature change is envisaged (separately for the monthly average maxima and minima, since diurnal range may change) to its existing site adjustments, if any. Calculation of E°days then proceeds as normal. In this way it is possible to trace, with fair accuracy, the best adapted grape maturity groups and wine styles under any climate change scenario, having reference to their optimum ripening conditions as set out in Table 3.1. Alternatively, the effects of progressive or permanent future temperature changes can be gauged approximately from sites having existing temperatures matching those projected. For greatest accuracy this requires that their latitudes and continentality be similar. Other factors such as sunshine hours, rainfall and relative humidity may of course also change somewhat, as will atmospheric CO_2 concentration; but primarily it is temperature that controls phenology and what is possible.

2.2   Temperature, growth and fruiting

Figure 2.1 in subsection 2.1.1 has already shown the generalized relation of vine dry matter increase to mean temperatures. From a base temperature of 10°C or a little less its rate reaches an optimum at around 22–25°C; but beyond that, unlike phenological development, the rate declines progressively to reach nil somewhere around 40°C mean. That is largely because accelerating assimilate losses by respiration overtake photosynthetic gains, which respond less to temperature. Canny (1973) cites evidence showing a virtually identical pattern for typical rate of solute translocation in temperate plants.

The fact that dry matter increase depends on photosynthesis, and hence on light as well as temperature, makes its temperature relations much more complex than those of phenological development. Unrelated factors that commonly limit photosynthesis include nutritional deficiencies and imbalances, water supply, diseases and pests, and light relations within the vine canopy.

Low temperature can directly inhibit photosynthesis. But usually, at these temperatures, it is other growth processes such as assimilate transport and cell division that limit first. That is especially so where nights are cold, because all growth processes other than photosynthesis continue at night. Consequent failure of the vine to use all the previous day's assimilate leads further to a temporary build-up, and 'back pressure' that reduces or delays the following morning's photosynthesis.

Two more factors complicate the relationship of temperature to growth and potential yield.

First, some acclimation to temperature can occur (Schultz 2000). For instance if temperatures rise slowly and remain consistently high, but within the limit of plant tolerance, the optimum for assimilation can move slightly upwards. Comparable limited acclimation is possible to steadily falling temperature. But whether these effects have any major impact on yield seems doubtful.

The second is almost certainly important in practice, particularly in hot climates. Whereas rate of phenological development is little influenced by canopy density and internal shading, that of assimilation can be so quite profoundly. Temperature responses of the latter, as reported in the literature, have typically been measured under good leaf and/or canopy illumination. But shading within the canopy creates an interaction with temperature that arguably lowers the optimum.

This can be reasoned from the fact that respiration by leaves continues regardless of their photosynthetic input. Shaded leaves that are hardly photo-synthesizing still respire at rates that increase with temperature, potentially to the point that they become parasitic, yellow, and fall. For semi-shaded leaves it is therefore advantageous for temperature to be low enough to minimize respiratory loss. Their optimum temperature for assimilation will be lower than under full illumination.

Optimum temperature of the canopy as a whole for assimilation, dry matter increase and fruiting will in theory therefore be lower in proportion to canopy congestion and internal shading. Cocks (1973) demonstrated such a relationship in the pasture legume *Trifolium subterraneum*, to the extent that the optimum temperature for sward growth fell from 20°C for spaced and well illuminated plants to only 10°C or less for the densest swards.

Problems arising from the combination of high temperatures with leafy canopies are common in the inland irrigation areas of south-eastern Australia, especially when further combined with very high fruit set and potential yields. Effects include delayed and uneven veraison and ripening, and poor fruit quality even if later temperatures are favourable. Coombe and Iland (2004) consider the quality implications of uneven ripening in some detail. See also Section 6.3.

Coombe (2000) reported an unrelated heat effect that can also contribute to uneven ripening. This appears to stem from direct heat damage to indi-vidual berries at some critical stage during veraison, which Coombe suggests may cause a temporary or permanent malfunction of the phloem feeding the berries. Recovery by some berries at various later times results in staggered ripening, with berries within clusters ranging in a scattered way from unaf-fected to still green and hard.

Later research by Soar et al. (2008) now suggests that this or some comparable phenomenon could have wider implications for wine quality than
hitherto recognized. These authors followed the detailed weather records
over 30 years for four contrasting Australian wine regions (Barossa Valley,
Coonawarra, Hunter Valley and Margaret River), focusing on sliding three-
week windows, one week apart, from approximately the end of berry cell
division to the latest harvest. Weather characteristics in each year, e.g. the
incidence of particular ranges of minimum and maximum temperatures,
were then matched against published red wine vintage ratings for each
region and year.

Adverse effects of sub-tropical rain and high humidity dominated quality
outcomes in the Hunter, as did the need for end-of-season sunny weather in
Coonawarra, which was the coolest region. But the Barossa and Margaret
River, both having intermediate temperatures and mostly dry summers,
showed a distinctive common response. Seasons with high quality ratings
typically had cool to mild weather just before and at veraison, whereas poorly
rated vintages had an above-average incidence of both high minima and high
maxima through the same physiological stage. Over that period Margaret
River also had on average high afternoon relative humidities in the good
years and low afternoon relative humidities in the poor years.

Soar et al. interpreted these results primarily in terms of the minimum
temperatures. But given the association of low vintage rating with high readings of both minimum and maximum temperatures, and the correlations
between the two, this seems unduly restrictive. Given further the associations
with afternoon relative humidities at Margaret River, a more straightforward
explanation seems likely. That is, that heatwaves with low relative humidities
around the beginning of veraison conduce to low wine quality, and mild,
equable weather over that period to high wine quality.

Simple heat scorch and berry shrivel in exposed clusters is also common
from veraison onwards in warm to hot climates, affecting only the cluster
parts immediately exposed. Experience suggests that the fruit is most susceptible to this form of damage at and soon after veraison, while sugar content is
still low. This also accords with the result of Downey et al. (2003), who found
that increase in berry flavonols, which protect against UV light, follows that
of anthocyanins and takes place in the latter half of ripening. Rising sugar
and flavonol contents bring evident late resistance to berry collapse, although
there can still be effects of heat on flavour ripening as will be discussed in
Section 2.3.

Not only heat causes uneven ripening. Several phenomena associated largely with low temperature during early spring growth and around fruit set can have comparable results.

A poorly defined temperature rise through spring can bring about staggered budbreak. This allows the earliest developing shoots to gain extra dominance in both time and vigour, a difference that tends to persist through setting, veraison and ripening. There appear also to be direct effects on the differentiating fruits themselves prior to setting, most probably around budbreak, that can lead to their later uneven development (Gray and Coombe 2009). Possible relationships to environmental conditions then remain to be explored. For a comprehensive review of inflorescence development from initiation to flowering, see May (2004).

Uneven setting and ripening can additionally follow from cold or wet conditions around flowering itself (Coombe and Iland 2004; Trought 2005). Two distinctive syndromes are recognized in addition to simple poor set. *Millerandage* ('hen and chicken') results directly from low air temperature at critical stages of flowering and setting, by interfering with pollen function and ovule development (Ebadi et al. 1996). Clusters then have varying proportions of small, seedless and poorly seeded berries that ripen before those with full seed complement and development. *Coulure*, in which clusters shed many small and undeveloped berries but the rest develop normally, results mainly from competition for assimilate from concurrent vegetative growth, typically under lush growing conditions with moderately cool temperatures, ample moisture and limited sunshine hours. Incidence varies with grape variety and its natural vigour, but if uneven within a vine or across a vineyard contributes to a staggered crop. The same conditions cause reduced and uneven fruitfulness in the lateral buds newly forming at the same time, that will grow into new shoots carrying the next year's crop. That in turn encourages further unbalanced growth and uneven ripening in susceptible varieties. All of these conditions (with the possible exception of *millerandage*, which can concentrate sugars and flavours) are seriously detrimental to wine quality, which ideally requires that all fruit should reach perfect ripeness more or less simultaneously.

We now turn to the other aspect of temperature that bears on wine quality and style: that during the middle and later stages of ripening.

2.3   Ripening temperatures

Whereas mainly early-season temperatures determine time of ripening, it is those during ripening itself that most directly influence grape and wine characteristics. I arbitrarily choose the last 30 days before an estimated average maturity date for making dry table wines to compare among environments or grape maturity groups in their ripening conditions. This covers roughly the period of engustment, or the build-up and conservation of flavour and aroma compounds in the berries. It also covers the period when the grapes' natural acid balance and sugar concentration at maturity are largely determined, together with tannin ripeness for red wines and in some varieties the persistence of 'green' compounds such as methoxypyrazines. Finally, it is a time when excessive rain or heat can greatly reduce quality (although around veraison is another critical time for heat damage).

As a rule, flavour development, and pigment development in red grapes, are favoured by intermediate temperatures similar to, or a little below, those found optimal for growth: i.e. about 20°C mean, with average maxima not above 26°C and average minima not below 14°C. Higher temperatures (especially maxima) bring no further increase in flavourant and pigment formation, and if much higher can accelerate their loss, as will be described below. Lower temperatures, while reducing flavourant and pigment formation, help to retain the more volatile aromatics. Low ripening temperatures, combined with high relative humidities as will be discussed in Section 3.3, also lead to flavour ripening at low sugar contents and therefore low wine alcohol contents, together with relatively high natural acids as appropriate to cool-climate wine styles. Thus grapes for delicate white wines may ripen best as low as 15°C mean temperature, as in the Rhine and Mosel valleys, and even lower for sparkling wines. Those for full-bodied (especially red) table wines appear to ripen best with mean temperatures around 18–22°C, as in Bordeaux and the Rhône Valley. Still higher temperatures with ample sun give the best sweet fortified wines, which need a high grape sugar content.

Little direct evidence is available on the evaporative losses of aromatic compounds from grapes, but a substantial literature exists on monoterpene losses from the leaves of forest trees. These studies show an exponential rise in losses averaging 10 to 12 per cent per 1°C rise in air temperature (Tingey et al. 1980, 1991; Lerdau et al. 1997; Staudt and Bertin 1998).

Such a steep response probably results only in part from increasing volatility with temperature. The permeability of plant cell membranes to solutes, necessary for their movement through cells to evaporative surfaces, increases some two to five-fold per 10°C rise in temperature (Collander 1959). The two mechanisms combined give ample potential for the steepness of the observed responses.

Evaporative loss of flavourants from grapes is complicated by another factor. It is now well established that many can conjugate with sugar molecules in the berry to form odourless and non-volatile glycosides that are possibly more resistant to chemical degradation as well as to evaporation. Their partial hydrolysis in winemaking and storage releases again the free and sensible aromatics. It may also be speculated, from research such as that of Wilson et al. (1984) and Strauss et al. (1987), that hydrolysis in the grape itself could contribute to the sudden rise in perceptible aromas and flavours late in ripening. Perhaps the glycosides and free flavourants remain in mutual equilibrium throughout, in which case factors such as low sugar supply (in over-cropping etc.) and highly evaporative day temperatures would tend to favour hydrolysis and loss of the free form at all ripening stages. An equilibrium of this type could help explain the finding of Marais et al. (2001), in South Africa, of a close inverse relationship between total monoterpenes in ripening Sauvignon Blanc grapes and average maximum temperatures for the week before sampling.

Another possible pathway to the same temperature relationship stems from degrees of fatty acid unsaturation in the berry membrane oils. As in plant tissues generally (e.g. Murata and Los 1997), lower temperatures generate reduced fatty acid saturation, a necessary mechanism for maintaining appropriate fluidity in the membranes. Unsaturation is also greatest in green grapes, and falls during ripening (Millan et al. 1992). On crushing, the fatty acids break down to shorter-chain derivatives that are a major source of herbaceous aromas and flavours in musts and wine. However, as in ripening apples and many other fruits (Mazliak 1970), they are capable of forming esters having apple-like and other fruity aromas: a process which, in the case of grapes, would presumably be largely during fermentation. The relationships to both temperature and grape maturity seem to fit a scenario in which herbaceousness and less ripe fruit characters from compounds of low boiling point are all correlated with degree of fatty acid unsaturation in the grapes.

It might be further speculated that fatty acid transformations during ripening could provide precursors of the cysteine-conjugated volatile thiols that dissociate during fermentation to give many of the distinctive aromas and flavours of Sauvignon Blanc wines, and contribute to those of wines from many other grape varieties (Tominaga et al. 2000; Murat 2005; Swiegers et al. 2005, 2006). This suggestion comes both from the chemical nature of the thiols and from the well-known relation between ripening temperature and Sauvignon Blanc aromatics.

A general conclusion on temperature is that no single mean or average can be laid down as universally optimal for wine grapes: whether for the growing season, for which grape varieties of differing maturities can be selected (Chapter 10), or for the final ripening period. Instead a range of optimum temperatures can be discerned, depending on grape variety and wine style. Table 3.1 summarizes those for the ripening period, along with suggested ripening-period optima for other climatic elements to be discussed in Chapter 3.

But Table 3.1 still gives only the broadest of ripening temperature guidelines. To gain a fuller understanding we have to look further to other temperature attributes, of which the most important is its patterns of variability: over the full year, at particular growth periods, and especially during ripening.

2.4   Temperature variability

2.4.1 Continentality

Continentality is the range between average mean temperatures of the hottest and coldest months. It is wide in highly 'continental' climates, such as in central and eastern Europe and central and eastern USA, and is accompanied by rapid temperature change in spring and autumn. Conversely it is narrow in 'maritime' climates, such as those of west coastal North America and most of the Southern Hemisphere, where seasonal changes are more gradual. Each climate type has its viticultural pros and cons, none of which is decisive as to overall advantage except that hot and cold continental climates tend to be ruled out by their respective summer and winter extremes.

Despite the lack of a simple relationship between continentality and suitability for viticulture, several aspects remain relevant to terroir.

The rapid spring temperature rise of continental climates provides a well-defined control over release from winter dormancy. This usually results in even budbreak, followed (conditions allowing) by even shoot growth, flowering, setting, veraison and ripening: an outcome clearly favourable for fruit and wine quality. Also the period of frost risk after budbreak is short, although reversion to winter conditions, if it occurs, can be extreme. Similarly the rapid autumn fall in temperature and daylength provides a clear signal for the cessation of vegetative growth and start of ripening, again contributing to even ripening. Against that the period of optimum temperatures for ripening is short, with potential for marked seasonal inconsistency of vintages. When all goes well, the coolness of late ripening and post ripening allows best retention of volatile flavourants, giving highly aromatic wines and favourable conditions for late-picked styles.

Maritime climates, usually at lower latitude, have opposite attributes. Seasonal extremes are less often limiting, allowing a wider spread of viticulture. In some highly maritime climates the lack of winter chill and gradual spring temperature rise can lead to incomplete winter dormancy and uneven budbreak. In compensation, rainfall allowing, there can be a long autumn period of optimum ripening conditions, giving flexibility in grape maturities and the economic advantage of an extended vintage. The best terroirs in such climates are often well suited to large-scale commercial viticulture. Wines tend to full-flavoured rather than delicate and aromatic.

The commonest overriding factor in continental climates is winter killing. Traditionally an average mean temperature for the coldest month of $-1°C$ has been regarded as the limit of reasonable safety for *Vitis vinifera* varieties (e.g. Prescott 1965; Branas 1974; Becker 1977; Jackson and Schuster 2001), although Branas notes that in Eastern Europe unprotected viticulture extends to areas with average coldest months as low as $-4°C$ with selection of sheltered mesoclimates and the most cold-tolerant grape varieties and rootstocks.

Unprotected winter survival of vines in such marginal conditions depends also on climates having low short-term temperature variability in the same way as escape from post-budbreak frost in spring: see subsection 2.4.2 below. A low variability index in winter indicates not only a relative lack of cold extremes for a given temperature mean, but also best chance for the vine to acclimate as temperatures fall in early winter. A low average Temperature Variability Index, or TVI (Gladstones 1992, discussed further below) for the non-growing season months gives a measure of this, and is included in the new-format viticultural climate tables exemplified in Appendix 2.

Viticulture beyond the limits described must usually depend on growing cold-tolerant North American *Vitis* species or their hybrids, of which the wine quality by and large is no longer acceptable; or else artificial winter burial of the vine crowns and canes, a practice now hardly economic. On the other hand reliable winter blanketing by snow, as in the lee of the North American Great Lakes, and use of cold-resisting non-*vinifera* rootstocks (see, for instance, Adams 1984, pp. 125-9) can facilitate quality viticulture in some cold-winter climates that would otherwise be ruled out.

2.4.2 *Within-season variability*

Within-season variability of temperature is a factor of macroclimate and mesoclimate unrelated to continentality, and has central significance for terroir. It comprises both variability between day and night, measured by diurnal range, and that occurring irregularly from day to day and week to week. The TVI is a generalized measure of this and is the sum, for any month, of the average diurnal temperature range and that between the average lowest minimum and average highest maximum. The best climates for producing quality wines have low variability. Their growing and ripening seasons are equable. The argument in its simplest form is that most temperature damage is done by the extremes. Irregularity of the extremes also reduces the capacity of the vines and fruit to acclimate to them.

The clearest example is that of spring frost damage. Regularly low spring mean and minimum temperatures cause the vine to adapt by delaying budbreak. The danger comes when temperature irregularity brings early warmth and growth, only to be followed by damaging frosts later. The Spring Frost Index, or SFI (Gladstones 2004) is the range, for any spring (or other) month, between its average mean temperature and its average lowest minimum. This gives a measure particularly of risk from advective frosts, caused by migrating air masses: i.e. the type of frost against which local site selection gives least protection. A low SFI gives not only the best chance of avoiding frosts after budbreak, but also of vines acclimating to them if they occur. Section 4.3 further discusses frosts in relation to topography.

Some acclimation probably also occurs in summer to high temperatures, with heat damage worst following mainly cool weather. The Heat Stressfulness Index, or HSI (Gladstones 2004) is a mirror image of the SFI. It measures the range, for any month, between its average mean and its average highest maximum. As already described, the period from just before to a little after veraison appears critical for such heat events, with even sub-clinical injury apparently able to compromise normal ripening and flavour development later. Abnormal heat stress during the final engustment period must also be important, whether for loss of aromatics or for berry shrivel and the appearance of 'dead fruit' aroma characteristics.

Gladstones (2004) defines frost risk levels according to SFI and maps both indices for Australian viticultural regions. SFIs <11 represent low frost risk, 11–13 medium risk and >13 high risk. Comparable values for HSI are <13 low heat stressfulness risk, 14–16 medium risk and >16 high risk. Note that the latter index must be considered in conjunction with, and adding to, the absolute values for the monthly average maxima.

Finally, it can be argued on theoretical and circumstantial grounds that even in the absence of extremes, a regularly narrow diurnal range during ripening should give best fruit and wine quality. Whereas daytime conditions govern photosynthesis and potential for heat damage, low temperatures at night limit the continuing processes of flavour biosynthesis and other non-sugar components of ripening. Volatile aromatics produced at night also have best chance of escaping evaporative loss and of conjugating with sugars etc. into more stable reserve forms.

The more rapid physiological (i.e. flavour as opposed to sugar) ripening that a narrow diurnal range makes possible has further advantages. Sugar contents at maturity are lowest, as are therefore wine alcohol contents. The shorter time available for loss of natural acids probably offsets acceleration of their loss due to warmer nights. Importantly, there is less opportunity for fruit quality loss through UV exposure, heat, rain, hail, disease, predators and simple senescence. The faster the formation of flavour and aroma compounds, the greater will be their capacity to accumulate relative to all sources of loss. Coombe and Iland (2004) discuss the question in some detail, and confirm that best fruit and wine quality almost invariably follows rapid flavour ripening.

Certainly ripening is slower at the low temperatures that are optimal for certain table wine styles. But what is desired there is the temperature of ripening, not the slowness. Within the limitations imposed by low temperature, flavour ripening still needs to be as fast and uninterrupted as possible. I argue that this is best achieved where temperatures are equable.

Whence, then, comes the widespread belief expressed in the popular literature, that wine quality results from a wide diurnal temperature range with cool or cold nights? I suggest it probably originated in the very hot environment of California's Central Valley. If day temperatures regularly reach close to 40°C, a night temperature falling to 20°C or less *seems* cool. It and the associated wide range are indeed essential if the fruit is to enjoy some period in the optimum range for production and preservation of flavours and pigments. However this is a feature of an extreme climate, outside the range of quality viticulture. It is true that parts of the Napa Valley also have unusually cool nights relative to their mostly optimal mean temperatures (see the Saint Helena climate table in Appendix 2). But this is to a degree offset by marked equability of their maximum temperatures. Also, highly valued parts of the lower valley benefit particularly from sea breezes that cool their afternoons, and night fogs that, if anything, tend to slow further temperature fall at night.

There is a further source of legitimate confusion that requires explanation. I am speaking here of *climatic* averages: that is, averages over the years and decades. But it also happens that at any one site the best ripening *weather* is fine weather, which will normally have a wider diurnal range than the site's average or that of unfavourably cloudy and rainy weather. A climatic ideal of narrow diurnal range may therefore seem counter-intuitive. It is nevertheless true. We can generalize: the best ripening weather at a site usually has a relatively wide range for that site, but within the context of a climate with an overall narrow range.

To these theoretical arguments may now be added the circumstantial evidence from Old World viticulture, as shown by the excellent maps of Johnson and Robinson (2001). Not only do the greatest viticultural regions have macroclimates with moderate to low within-season temperature variability as measured by the overall Temperature Variability Index. Within them, the acknowledged greatest vineyards are almost always in the topographic and mesoclimatic situations with best night air drainage, and therefore narrowest diurnal temperature range.

At the cool limit of viticulture a low diurnal range and low variability of the minima are in fact essential if grapes are to be grown and ripened at all. Only then can the vines escape spring frosts; while during ripening it is *relatively* warm nights that enable ripening with the cool day temperatures needed to preserve delicate aromas.

Given that the locally highest temperature means stem mainly from high minima, that explains the European saying that the best wines come from the warmest sites in the coolest climates.

But there is a further paradox to be explained: that on average the best wines also come from the warmest seasons, when the grapes ripen early and under often quite high mean temperatures. That seems final proof to me that the primary factor for quality during ripening is not so much coolness as equability. Equable warmth (to a point) enhances flavour and colour intensity, without too great a loss of the more volatile aromatics.

Chapter 3

Other Climate Components

3.1   Light intensity and exposure

A common New World assumption is that unlimited bright sunshine is beneficial for viticulture. Recent findings have challenged this idea.

Directly exposed leaves of most temperate plants need only about a third of bright sunshine intensity to photosynthesize at maximum rate. In fact photosynthesis can be inhibited when intensity greatly exceeds that, especially if accompanied by high temperatures (Iacono and Sommer 1996). Moreover, leaves continuously exposed at full photosynthetic levels soon become assimilate-saturated, leading to a closing down of photosynthesis in the middle of the day. Low-intensity or intermittent sunlight, such as under fog or broken cloud, can therefore allow just as much photosynthesis and occasionally more, since diffuse sources allow more penetration of direct light within the canopy (Allen et al. 1974).

The relevance of these relationships to terroir is that climates with either low sunlight intensity, as at high latitude, or with much light or broken cloud, can combine largely undiminished assimilation with least heating of exposed fruit. Berries exposed to bright sunlight for long periods commonly reach 12–15°C above air temperature (e.g. Smart and Sinclair 1976). This may help ripening at its cold limit, but is more likely to be detrimental elsewhere. Adverse effects include clinical heat injury and, sub-clinically, the degradation or evaporative losses of pigments and aromatics as described in Chapter 2. It can also lead to excessive influx and berry concentration of sugar, as will be discussed in Section 3.3.

Careful canopy management in hot, sunny climates can to some extent reduce these effects. We now know, from some outstanding recent Californian research (Bergqvist et al. 2001, and especially Spayd et al. 2002), that damage to exposed fruit relates directly to the temperature it reaches, not to light as such. The 'dappled light' provided by moderate canopy cover can replicate the effect of broken cloud in allowing enough direct illumination of the fruit for best ripening biochemistry, while avoiding prolonged cluster exposure and cumulative heating. Also vine rows can be oriented other than north-south to confine direct lateral illumination to the mornings and late afternoons when air temperatures are moderate.

This can be achieved by orienting vertically-trained rows north-east to south-west in the Northern Hemisphere, or north-west to south-east in the Southern Hemisphere. In very warm and sunny climates an east-west row direction may be best, allowing canopy trimming and fruit exposure to mainly indirect light on the side facing away from the sun, but maximum protection with only limited intermittent sunlight exposure of the fruit opposite. Alternatively an uncontrolled or bush-type canopy gives satisfactory protection, as shown by Wolf et al. (2003) in South Australia's Barossa Valley and used traditionally in the Mediterranean region.

At the cool and cloudy limit of ripening, on the other hand, the standard advice of north-south rows remains valid. Bilateral leaf trimming in the fruit zone, or pruning systems such as the Geneva Double Curtain that expose fruit at the top of the canopy, then give the fruit maximum illumination and heating as needed for ripening under cool and often low-light conditions.

We can conclude that for table wines some degree of cloudiness is desirable more or less universally. In cool climates it contributes to the temperature equability needed for spring frost avoidance, and during ripening to retention of the highly volatile berry aromatics essential for cool-climate wine styles. In warm to hot climates there is still often a need for spring frost avoidance, while during veraison and ripening cloud helps to minimise fruit heat injury and loss of surviving aromatics and pigments.

The subject of cloudiness cannot, of course, be dissociated from that of rainfall. A good terroir combines enough (preferably high or broken) cloud or fog with minimal rainfall at the critical stages of fruit-set, ripening and harvest. The anomalous combination is one reason why the best terroirs are climatically so specialized.

3.2   Rainfall

No universal guideline is possible for total rainfall requirement. This varies with temperature and related factors, while in many regions any deficiency can be made good by irrigation. On the other hand there is now a substantial consensus on what moisture regimes (by whatever pathway) give best vine balance and wine quality. It can be summarized as follows.

1. In winter-spring, enough rain or snow to replenish water reserves throughout the soil and subsoil without waterlogging. (This also depends on soil type and depth.) Reserves and late-spring rainfall ideally supply enough moisture for unrestricted spring growth, flowering and setting.
2. During flowering and setting, no rain and plenty of warm sunshine.
3. Between setting and veraison, moderate moisture stress to halt further shoot extension and branching. The aim is to establish an open canopy that allows intermittent direct illumination of the fruit and newly-forming buds. The vines become hardened against further stresses, while restricted vegetation helps to insure against them by limiting later water needs. For red wines, stress that starts soon after setting can help quality by restricting berry size.
4. From veraison to harvest, mild to moderate stress only: enough to maintain the inhibition of vegetative growth and encourage root growth, but not so much as to cause leaf loss or interfere with photo-synthesis and other leaf functions. There should be no hail, and no heavy rain late in ripening or at harvest.
5. After harvest, enough moisture to maintain leaves and continue assimi-lation during any remaining part of the growing season, for build-up of reserves and proper cane and bud maturation.

Clearly few, if any, climates meet all these criteria, although some seasons in the best of them do. Seeking the best terroirs in respect to rainfall and irrigation is always a compromise, which must also take into account the commercial purpose of the viticulture. Leaving aside bulk wine production in hot, dry areas with full irrigation, however, two main types of rainfall regime come into focus for quality wine production.

The first is that of seasonally uniform rainfall in mild to cool climates, as in western and central Europe, New Zealand and Tasmania; and, with reservations about winter killing, parts of eastern Europe and north-eastern North America. In these mostly damp and cloudy climates, the best terroirs are generally those in the rain shadows of hill or mountain ranges with regionally the least rainfall and most sunshine, giving enough moisture stress in summer-autumn to contain vigour and give a reasonable chance of dry ripening and harvest.

The second is in mild to warm mediterranean or sub-mediterranean climates, with ample winter-spring rainfall but dry summers. In such climates the adequacy of late-season moisture depends either on good root-accessible soil reserves or supplementary watering as needed.

Traditional quality viticulture in Europe conforms to the first climate type. It is the second, in the Mediterranean region itself but notably also west-coastal North America, Chile, Australia and South Africa, that now challenges Old World viticulture in producing high-quality wines.

But temperature, light and rainfall patterns are not the only climatic factors shaping terroir. A fourth, hitherto largely ignored, must now be considered.

3.3   Atmospheric humidity

Early New World opinion generally held that a dry atmosphere, like cloud-less summer skies, was an advantage for viticulture. That is certainly so for disease control, and to that extent, for wine quality. But historical experi-ence and some recent research findings now suggest that opposing factors outweigh this, at least for table wines and where diseases can be adequately controlled.

One is that low relative humidities (and associated lack of cloud) correlate causally with wide temperature variability, and hence with enhanced risk of killing frosts and summer heat stress. Even without these, such a tempera-ture regime is arguably unfavourable for ripening (subsection 2.4.2).

Second, daytime low humidity and high temperatures accelerate leaf transpiration of water, both absolutely and relative to the simultaneous intake of carbon dioxide. The vine uses more water for the same photosynthesis, and with more risk of damaging moisture stress.

Perhaps most importantly, there is now evidence that sugar and other solutes move in phloem sap into the ripening berries in direct response to evaporative draw from the berry surfaces (Rebucci et al. 1997; Dreier et al. 2000). Low humidities and high temperatures thus accelerate berry accumu-lation of sugar, and presumably of other phloem solutes including potassium. This is so both absolutely and relative to flavour and tannin ripening, which strong stresses if anything impede. Added to this, extreme evaporative draw and soil drought may cause withdrawal of water from the fruit back into the vine, further leading to passive sugar concentration in the berries. The find-ings fit the common field observation of very rapid sugar increase during hot, dry spells at ripening versus a near-cessation of increase when the weather is cool, cloudy and damp; also, why grapes in warm, dry, sunny ripening climates typically reach high sugar levels before flavour and tannin ripeness.

Ideally all components of ripening should reach their desired end points – for given wine styles – simultaneously. Rapid sugaring may be appropriate for sweet fortified wines, for which indeed some further sugar concentration by raisining can be desirable. But for balanced table wines the sugar content at full flavour maturity is at best only moderate, which requires a less evaporative climate during ripening. This is a matter of both temperature and atmospheric relative humidity, or as the combination is expressed directly, saturation deficit.

It is worth underlining that relative humidity and saturation deficit, like cloudiness, correlate only partly with rainfall. Some coastal regions can have predominantly high relative humidities but almost complete summer drought, as along the immediate west coasts of both North and South America and the south-western and southern coasts of mainland Australia and South Africa. This is a further reason why good viticultural climates are specialized and rare, since enough humidity tends elsewhere to be correlated with too much rain.

Relative humidities or saturation deficits measured in the mid to late afternoon are those most related to evaporative water loss. Morning relative humidities are high in most climates and do not differ greatly across viticultural regions.

Mid-afternoon relative humidities, together with maximum or afternoon temperatures, are thus a valuable indicator for viticultural terroirs. Table 3.1 (Section 3.5) gives values suggested to be optimal for individual wine styles.

3.4   Wind

The ill effects on vines of strong wind exposure are obvious and need no comment here. Even moderate exposure to cool winds in spring can reduce vine vigour and yield (Dry and Botting 1993).

Favourable effects of moderate winds, on the other hand, are less obvious but no less real, and need to be considered as potentially positive parts of terroir. For instance open exposure to wind typically reduces radiative frost risk, although it can increase that of advective frosts (see Section 4.3). Within the vine canopy, moderate wind helps to prevent the build-up of excessive humidity and hence diseases. It also reduces day-time depletion of atmospheric carbon dioxide within the vine canopy (Allen 1971). Leaf movement increases intermittent sunlight penetration to leaves below. Such dappled light is used with high efficiency for photosynthesis (Kriedemann et al. 1973), while giving beneficial fruit exposure without over-heating. For fully exposed and sun-heated fruit, winds are nearly always cooling.

Of further significance for terroir are regional winds and sea or lake breezes, as shaped by the dispositions of land, oceans, lakes and hill or mountain ranges. Proximity to oceans (and on lesser scales to lakes or major rivers)

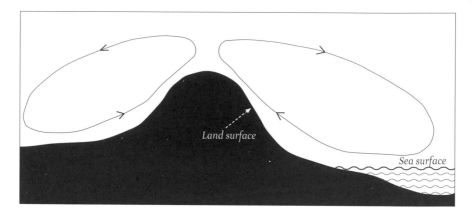

Figure 3.1. Generalized pattern of afternoon air convections on seaward and inland slopes of coastal ranges.

promotes equability of both temperature and humidity, through a convective cycle that brings alternating land breezes in the mornings, and cool, humid sea or lake breezes in the afternoons and early evenings as shown in the right-hand half of Figure 3.1. The cycle intensifies where coasts or lakes are backed by hill or mountain ranges with penetrating valleys, since typical valley winds flow down-valley in the mornings and up-valley in the after-noons and evenings. Afternoon up-valley winds are still more intense and prolonged where the backing slopes receive direct heating from the midday and afternoon sun.

Land/sea convections are critical in moderating the coastal summers of continents with hot, dry interiors. Notable examples in Australia are the Swan and Hunter valleys on the west and east coasts respectively; while along the south coast, strong daytime heating of the continental interior reinforces the normal summer south-east trade-winds, resulting in mostly regular cooling afternoon winds and mild viticultural climates for some distance inland, albeit with high temperature variability factors.

Sea breezes are also important along the west coasts of both North and South America. But whereas Australia's relatively warm oceans produce opti-mally mild winds, those off the American west coasts are cold, presenting sharp contrasts with the immediate interior. On-shore winds there are also blocked by coastal ranges with only a few penetrating valleys. The result is much coastal fog, with distance and inland direction of the coastal valleys the major factors determining summer climates. These can change rapidly and crucially for viticulture over short distances, depending on local topo-graphy. Halliday (1993) and Swinchatt and Howell (2004) illustrate this for

California's coastal valleys. Their terroirs are complex and highly localized as a result of this and some equally complex geology (Swinchatt and Howell 2004).

The effects of such winds on vine phenology can be expected to depend on their temperatures. In a warm to hot climate with relatively warm sea breezes they probably make little difference, since the breezes are themselves at or close to temperature saturation for rate of phenological development. In a cooler climate, however, the effects can be significant. A cool sea breeze arriving at, say, 2 p.m. reduces temperature rapidly and holds it down for the rest of the afternoon. Given that the day's maximum temperature may normally be reached by then, the breeze will have little effect on the recorded maximum or daily mean; but effectively the day is cooler. In my observation of climates and vine phenologies in south-coastal regions of Western Australia, this can retard phenology in only moderately cool climates. In very cool climates, and where marine winds come from cold oceans, the limitations on phenology are commonly critical. Here viticulture is usually possible only on the leeward slopes of hills or mountains that protect from cold winds, as will be described later in this section.

The suggested temperature adjustments for wind exposure (Table 2.2) are subject to a rough automatic control for these differences. Capping each month's effective mean temperature at 19°C fully discounts cool wind effects in all months warmer than that. But as the number of growing season months below 19°C increases, so does the impact of the adjustment on seasonal E°days.

More important in warm to hot climates is the direct effect of winds on heating or cooling of the ripening fruit. Heating of the clusters, whether sun-exposed or not, is a cumulative process. A peak of sun or air temperature can be tolerated if soon relieved, whereas sustained high temperatures may not. Cool sea or lake breezes arriving in mid or even late afternoon can therefore still have large benefits. Moreover the breezes are humid, as well as cool. The combination means that evaporation from the berries is reduced, and with it presumably the excessive berry sugar accumulation that is typical of hot, dry climates.

Sea and lake breezes, then, can play a far greater role in permitting quality viticulture in otherwise hot, sunny climates than their effects on standard climate statistics would indicate. They are a major positive component of terroir in these situations.

The wind systems associated with the Great Lakes in North America, and with medium to large lakes generally in cool, continental climates, have a unique and viticulturally positive role of their own. Their coldness in spring retards budbreak in vines within their spheres of air convection, while at the same time usually advancing the period of freedom from serious spring frosts. Conversely at the end of the growing season their accumulated warmth helps to extend ripening. See Kopec (1967). Ripening conditions are then equable and fairly humid: ideal (provided autumn rains hold off) for late-picked styles and ice wines such as have become a Canadian specialty. Moisture from such lakes also increases winter snowfall on their lee sides, enhancing protective winter snow cover for vines there.

We can now consider what happens on the reverse slopes of the hill and mountain ranges that protect from cool and cold winds. Broadly this constitutes a mirror image of what happens on the exposed sides. Up-valley afternoon and evening convections draw air from further inland or (usually) lower latitudes (Figure 3.1). The convections are strongest where the slopes face the midday or afternoon sun. Instead of cooling, the effect is one of prolonging afternoon and evening warmth beyond normal expectation from the recorded maximum temperatures. Adjustments to the maxima to represent effective afternoon temperatures are therefore upwards.

The viticultural areas of Central Otago, in the south of New Zealand's South Island, provide a likely example, with surrounding mountain slopes to the east, south and west. Mapping of sea-level growing season isotherms for the South Island (Figure 3.2) shows here a steep warming trend to the north, so that up-slope afternoon-evening convections are warmed not only by contact with sun-heated, north-facing slopes, but also by warm air drawn in from the north. This helps to explain why Central Otago can ripen grapes despite a climate that nominally is very marginal. A similar situation exists on a smaller scale in the Waipara region on the east coast north of Christchurch. Here coastal hills protect from the cool marine winds that otherwise sweep across the Canterbury Plains, while their inland, north-west-facing slopes both enjoy direct heating and draw in effectively warmer air from further inland. It explains why Waipara is viticulturally much warmer than the plains just to the south, and more closely comparable to the Marlborough region further north.

Similar factors probably warm the growing seasons of southern slopes and approaches of the various east-west mountain ranges in central and eastern Europe, as well as helping to protect their vines from killing by polar air masses that periodically migrate west and south in winter (see Section 4.3 on frosts).

The temperature adjustments for wind exposure are more speculative

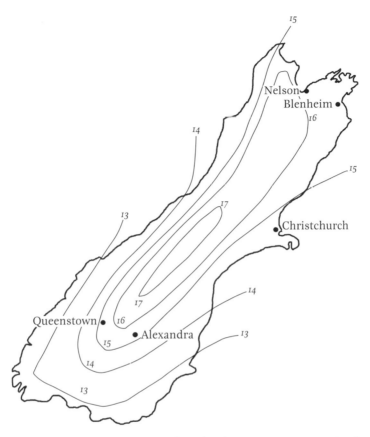

Figure 3.2. Growing-season averages of sea-level mean-temperature isotherms for the South Island of New Zealand. Based on climate data of WMO (1998) and NIWA (2003).

than most, and should probably be included only where the need for them is clearly apparent. It is also worth noting that because they are applied only to the maximum temperatures, their influence on E°days is much smaller than those of adjustments to the minima, as the following example shows.

A rise in the minimum of, say, 1.0° raises the mean by 0.5°; but the narrowing of diurnal temperature range by 1.0° increases the *effective* mean by a further 0.25° for those months to which the adjustment applies. A comparable rise in the maximum, however, widens the diurnal range by 1.0° and hence reduces the increase in effective mean by 0.25°. The effect on E°days is only one third of that stemming from a similar change in the minimum. This conforms to the thesis that rate of phenological development is limited primarily by night temperatures.

The main implications of wind adjustments, then, are for temperature variability and the incidence and severity of heat extremes. As such they are significant for terroir despite their small effects on estimated ripening capacity.

3.5   Ripening-period ideals for wine styles

Not all climatic elements lend themselves to the convenient definition of ripening-period ideals. But generally the more important ones do, with standard climate records (or derivatives of them) available that give reasonable pointers to impacts on the vines and fruit. They include monthly averages of temperature means, and of average and extreme minima and maxima; various indices of temperature variability (Section 2.4); measures of average daily sunshine hours and cloud cover (Section 3.1); rainfall (Section 3.2); and afternoon relative humidity (Section 3.3). Taken together these provide a conspectus of viticultural climate for any growing-season month; and once an average maturity date is known or estimated for any combination of grape variety and climate, of that for the critical ripening period.

The guidelines shown in Table 3.1 are a synthesis of observed ripening conditions across the world's established great viticultural regions for broadly defined wine styles. They are not totally inflexible. For instance some grape varieties have unique adaptations outside the general run for their maturities,

Table 3.1. Suggested optimal climate averages for the final 30 days of ripening[1] for different wine styles. From Gladstones (2004).

	Sparkling wines	Table wines			Fortified wines
		Light	Medium–full	Full[2]	
Average mean °C	12–15	15–18	18–21	21–24	21–24
Average max. °C	<21	<24	<27	<30	<33
Av. highest max. °C	<27	<30	<33	<36	<39
TVI[3]	<30	<33	<36	<39	<42
Sunshine hours/day	5–6	6–7	7–8	8–9	>9
Rainfall mm	<50	<50	<40	<30	<20
Early p.m. R.H.%	60–65	60–65	55–60	50–55	40–50

1. Final 30 days to estimated average date of maturity for dry table wines.
2. Best in coastal or other suitably equable climates, elsewhere mostly bulk wines under irrigation.
3. Temperature Variability Index: see subsection 2.4.2.

or may be able to make a range of valid wine styles across diverse ripening conditions. Needs differ depending on commercial purpose: one case most requires freedom from viticultural hazards and reliable vintages; another accepts some vintage failures for the sake of outstanding wines in the best seasons. Coastal afternoon sea breezes can moderate climates that mean temperatures indicate to be hot, as described in Section 3.4. Similarly, canopy management can to some extent modify fruit microclimate and hence wine style. But while all of these provide significant departures, they do not invalidate an overall correlation between climate measurements and potential wine style. That still gives a useful guide for evaluating new environments for viticulture, selecting grape varieties most likely to succeed in them, and forecasting the wine styles they will most naturally produce.

Chapter 4

Geography, Topography and Soil

4.1 Latitude

High and low-latitude vines have substantially different light and, to a lesser extent, temperature regimes. Summer days at high latitudes are long but sunlight intensities low, due to the oblique angle of the sun. Mostly there is also more cloud than at low latitudes, and because of the sun angle the clouds cast more extensive shadows. Light intensity through the grape-growing season nevertheless remains much of the time at or close to that needed for maximum photosynthesis. Total seasonal assimilation can there-fore be greater at high than at low latitudes with comparable temperatures, as Huglin (1978, 1983) and other authors have argued. Most of this difference accrues in mid-summer, when high-latitude vines not only enjoy very long days, but are also unlikely to be stressed by excessive heat, total or ultra-violet light intensities, or (as a rule) low humidities or water supply. Superior assimilate build-up then assures enough sugar supply to ripen moderate crops well into the autumn, despite fast-diminishing temperatures and sunlight; and usually, for a surplus to remain to support early growth in spring even if there is little further assimilation after harvest.

That the sun is low above the horizon during ripening at high latitudes has further implications. Combined as it commonly is with cloud or mist and high humidities, this results in very restricted daily temperature amplitudes and low measures of all temperature variability indices. With continuing fine weather the conditions are ideal for engustment and retention of volatile aromatics and natural acids, and the berries reach full flavour ripeness at low sugar content.

All these factors agree with historical experience that wines of delicacy, aroma and moderate to low alcohol are most naturally produced at high latitudes. Precarious ripening in many such climates means, however, that vintages are variable and failure common. Commercial success depends on very high quality in successful vintages and prices to match.

Low-latitude terroirs have generally opposite characteristics. Near-vertical midday sun in summer and the sharper change from day to night lead to greater short-term temperature variability, other than in sub-tropical climates where summer rain and humidity are limiting hazards. Dry-summer medi-terranean-type climates, by contrast, tend to excess light intensity (including ultra-violet) and temperature during veraison and sometimes ripening. Both can be favourably moderated by canopy management, and in near-coastal areas by afternoon sea breezes and cloud. Natural wine styles are full-bodied rather than subtle, although lighter body and good complexity are attainable with enough afternoon cloud and humidity, producing wines comparable to those of high latitudes but often more reliably.

4.2   Altitude

Is it an accident of history that most of the world's great wine-producing regions are close to sea level? Few are much above 500 metres and the great majority are at 300 metres or less (see Johnson and Robinson 2001). This applies in part even to hot regions where high elevation might be thought essential.

Earlier (Gladstones 1992) I suggested that sites above about 500 metres might be disadvantaged because of lower carbon dioxide concentrations compared with those at sea level. With increasing altitude less carbon dioxide is taken into the leaves for the same transpiration of water, resulting in reduced water efficiency and more risk of moisture stresses. However, this now seems unlikely to be a major factor, given steadily increasing atmo-spheric carbon dioxide.

Perhaps more important is diminishing atmospheric turbidity with alti-tude (Geiger 1966). Sunlight becomes more intense and less diffuse, leading to more risk of excessive intensity for directly illuminated leaves and clusters, and less diffuse light penetration within the canopy. Moreover the proportion of energy as ultra-violet radiation increases with altitude. At low latitudes this adds to an already high UV intensity due to short passage of the near-vertical sunlight through the ozone layer.

Cumulatively, these factors probably do have significant effects. Combining them with others of latitude *per se*, it could be generalized that desired seasonal and ripening-period temperatures are better achieved by low sites at high latitudes than by high sites at low latitudes. Undoubtedly there is some advantage in otherwise hot, low-latitude regions from choosing highland sites with more optimal temperatures; but these probably cannot fully match environments with the same temperatures at lower elevations in higher latitudes. Best at low latitudes are coastal and near-coastal terroirs, where marine influences result in cooling, humidifying and cloud at low elevations so as to combine some of the benefits of high latitude with generally more reliable ripening. History does not look to have been mistaken.

4.3   Topography, air drainage and frosts

The contributions of topography to above-ground climate are well known and uncontroversial: see Geiger (1966). I treat them in some detail in *Viticulture and Environment* (1992). Geiger also covers the highly important subject of below-ground climate, which forms part of our Chapter 5 discussions on soils and root systems.

The largest climatic effect of topography, apart from that simply of altitude on temperature, is on cold air drainage at night, and hence local susceptibility to autumn, winter and spring frosts. Other effects related to slope and aspect will be discussed in the following section.

Two types of frost are normally distinguished: advective frosts brought by the migration of cold (usually sub-polar) air masses to lower latitudes under certain synoptic conditions, and radiative frosts caused by local radiative heat loss from the ground or vegetative surfaces under clear night-time skies and still conditions. The risk of vine damage from either type of event can be largely estimated from the day-to-day and week to week variability of minimum temperatures as measured by the Spring Frost Index, or SFI. See subsection 2.4.2 .

Advective frosts are influenced chiefly by major topographic features that channel or obstruct air-mass movement. For instance the unimpeded flow of arctic air down the Mississippi/Missouri drainage basin in the USA can carry extreme winter cold as far south as the Gulf of Mexico; however west coastal regions are protected by its containment behind the Rocky Mountains, while to a lesser degree the east coast shelters behind the Appalachian Range.

Cashman (2000) describes a typical advective frost event in south-eastern Australia. Here sub-polar air from the south may be further chilled by radiative cooling as it perches over the Australian Alps before spilling inland. Frost on the inland slopes and adjacent plains then depends on how medium-sized topographic features assist or obstruct its drainage. For instance the Whitlands High Plateaux region in Victoria is on a north-projecting ridge of the Alps, and partly escapes because the coldest air sinks and by-passes it down the valleys of the King River and Boggy Creek on either side.

However, the risk beyond is increased in line with the efflux from the valleys. The vineyards of Orange, New South Wales, are similarly protected by being on the north-facing lee of Mount Canobolas, around which cold air preferentially flows down the valleys of the Lachlan and Macquarie rivers.

Small hills and ridges can still have some influence on exposure to winds bringing advective frosts; but their greater significance is for radiative frosts. Under otherwise still conditions, air cooled and made denser by contact with land surfaces chilled at night by radiative heat loss trickles downwards via drainage channels to collect in depressions or over flat land below. This down-hill surface movement sets up a convection which draws down unchilled air from above, warming the parts of the slope above the channelled or settled cold air. Resulting 'thermal zones' (Geiger 1966) are mostly on the middle and lower slopes but are specially pronounced on those of isolated hills and projecting ridges, which have little external source of cold air by drainage. They have distinctly higher night and minimum temperatures than the depressions, valley floors or flat land, whereas their day and maximum temperatures are unaffected except to the extent of the normal lapse rate with altitude. The result is higher mean and average temperatures, with a marked reduction in diurnal temperature range and incidence of frost. Where the cold air flows away freely, for instance from a scarp onto a plain, or down unobstructed valleys, the thermal zone descends to the gently rising benchlands that exist at the foot of many slopes and along the sides of valleys.

Experience has shown that such thermal zones produce the best wines: see, for instance, the maps in Johnson and Robinson (2001) for Burgundy, Chablis, Alsace and Champagne. But important and often correlated factors of soil type and drainage are usually involved as well, together with aspect and steepness of slope.

4.4   Aspect and slope

The influence of aspect on effective temperature works jointly with those of latitude, steepness of slope, and time of day and year. Despite the complex trigonometry this entails, the principle for present purposes is simple. On flat ground the intensity of surface insolation at any time on a cloudless day is proportional to the sine of the sun angle above the horizon. (Greater loss due to a longer passage through the atmosphere at low angles somewhat intensifies this relationship.) Aspect determines whether a slope faces predominantly towards the midday sun (increased insolation) or away from it (reduced insolation). Steepness of slope in turn governs the relationship according to the sine of the sum of the sun angle above the horizon and that of a slope facing it, or of the difference in angles where the slope is facing away.

The temperature adjustment guidelines in Table 2.2 reflect these relations. Both aspect and slope have their most critical effects at high latitudes, due to low sun angle and the fact that most temperatures are in the limiting range for vine phenological development. The effects diminish to negligible at the lowest viticultural latitudes where the growing-season midday sun is close to overhead and much of the season the temperatures are non-limiting.

The adjustments in Table 2.2 are listed as plus or minus: plus for slopes facing the midday sun, and minus for those facing away. They converge to nil for east and west aspects. Full positive adjustments can be made where slopes face approximately southeast to southwest in the Northern Hemisphere or northeast to northwest in the Southern Hemisphere.

Viticultural lore holds firmly that east-tending aspects are best for wine grapes. Good reasons exist to support this belief.

The first is that excessive rain and cloud come mostly with weather systems from the west. (East coasts are partial exceptions to this.) In European high-latitude climates these are primary limitations of viticulture. The sheltered eastern slopes of mountain ranges, with adiabatic warming of westerly winds as they descend, are the warmest and sunniest of their regions but without the often excessive heat and dryness of comparable topographies at low latitude. Alsace and Burgundy in France, the Rhine Valley in Germany, the Willamette Valley in Oregon, and all viticultural regions of the South Island of New Zealand are good examples.

Second, an easterly aspect has earliest exposure to the morning sun, warming both vine and soil at the time of day when temperatures are lowest and most limiting. Photosynthesis can start and accelerate early, while fruit can gain all the benefits of sun exposure without over-heating. By contrast a westerly exposure maximizes heating when air temperature is already at its highest and most likely to cause injury. An easterly aspect promotes effective temperature equability, a westerly aspect detracts from it. This difference may, however, be moderated in west-coastal areas where a westerly aspect catches afternoon sea breezes.

4.5   Soil and above-ground microclimate

The daytime accession of sunlight to the soil surface, and its reflection or absorption and later re-radiation as long-wave heat, is an aspect of terroir that recent research has too much neglected. Aerial measurement and mapping of night-time long-wave radiation has indeed long been used in Germany and elsewhere, to evaluate sites for frost risk and overall warmth (Becker 1977), but good reasons exist to think that the implications are wider. It is easy to appreciate that heat absorbed by sun-facing, rocky slate soils of the Mosel Valley warms the vines and fruit at night and allows Riesling to ripen in a climate that otherwise would be too cold. But why do very rocky soils that do much the same epitomize the best of Châteauneuf du Pape, in France's warm southern Rhône Valley; or likewise the sun-facing rocky terraces of Portugal's hot Douro Valley?

An apparently universal superiority of stony or rocky soils for quality viticulture, regardless of climate, has been acknowledged for centuries: e.g. Rendu (1857), Petit-Lafitte (1868) and Portes and Ruyssen (1886). More must be involved than merely extra heat to enable ripening.

One factor is undoubtedly that a stony or rocky surface helps to protect against erosion, making viticulture possible on mesoclimatically desirable slopes that would otherwise be too steep. Yet the same quality advantage of stony soils extends to many flat locations as well.

This section will concentrate on the influence of soil characteristics on microclimate above the ground. Geiger (1966) and van Eimern (1968) treat the subject in detail. Chapter 5 will tackle the equally (perhaps more) important subject of microclimate below the surface, and its postulated effects on the root system and fruit ripening.

Two major factors influence the energy balance between soil surface and the air and vines above it.

The first, and more problematic, is soil surface colour. A light-coloured surface reflects a material part of incoming light energy, with a correspondingly reduced proportion absorbed and converted to soil heat. Net consequences for the vine and fruit are mixed. Extra light to the shaded parts of the vine is an advantage for photosynthesis and probably for the ripening berries. On the other hand in warm and hot climates it can mean a harmfully enhanced energy load through the hottest times of the day. Dark surfaces absorb more of the energy (depending also on their thermal conductivity and heat storage capacity, as discussed below) and re-radiate most of it later when temperatures are cooler. The daytime difference may not be critical for white grapes, which themselves reflect away a fair proportion of incident light, but can be so for red grapes. Their red pigments absorb much of the light and convert it to heat, adding to their heat load. Nor do berries cool by transpiration as can leaves, as they have very few stomata. Perhaps these are reasons for the widely believed association, at least in warm to hot areas, of white soils for white grapes and red soils for red grapes. Perhaps also the spectral quality of the reflected light plays a role.

More certainly dark-surfaced soils that reflect little light and convert nearly all to heat, to be re-radiated later under cloud or in the evening, are regarded as a general factor for successful viticulture in cool climates where night temperature most clearly limits ripening. They warm the evenings and reduce temperature variability. Examples here include the dark slates of the Mosel, and the 'bituminous' soils of Württemberg and of the Meuse Valley in southern Belgium.

The complementary second factor, which is critical in the examples just cited, combines a soil's heat conductivity and its volume capacity for storing heat. A low value of each means that the surface heats shallowly to high temperature and re-radiates its heat intensely and more or less immediately. This is clearly undesirable. A conductive soil with good heat storage capacity, on the other hand, absorbs to depth and retains most heat with least rise in surface temperature. The stored heat is re-radiated more steadily and over a longer period. This both benefits night-time ripening processes and reduces the risk from morning frosts.

Solid rock conducts heat many times faster than still water, although its volume heat storage capacity is only about half. Still water in turn conducts many times faster than still air, which is a very effective insulator. So is organic matter, especially if loose and dry. A stony or rocky soil with relatively little organic matter absorbs, conducts and stores most heat during sunshine and returns most at other times. Moderate dampness of the underlying fine soil (maintained in part because surface stones and rocks block evaporation) helps in soil heat storage through increasing both conductivity and volume heat storage capacity. Moderate soil compaction helps further through a reduction of insulating air spaces.

All these factors, singly or in combination, improve vine temperature relations above ground. The resulting greater temperature equability benefits the vine and fruit quality across *all* viticultural climates.

Temperature adjustments for soil type, as in Table 2.2, can at best be approximate, if only because soils vary so much over short distances and with depth. Nevertheless they play an essential part in improving the estimation of E°days.

Soil conditions at official recording stations can reasonably be taken as neutral, short of evidence to the contrary (e.g. the presence of underlying or nearby paving etc.) Nor will soils chosen for quality viticulture often warrant a negative adjustment. Towards the cool limit of viticulture most will neces-

Pergola / raise fruit some / raise heat

sarily have characteristics requiring a significant positive adjustment.

Besides the qualities of vineyard soils themselves, one further factor beyond those noted in Table 2.2 deserves consideration. That is the nature of land surfaces up-slope from vineyards, especially where water bodies below them set up or reinforce day/night air convections. Vineyards described and illustrated by Halliday (1991) and Cooper (1993) in Central Otago, New Zealand, are a good example. Dark, north-facing rubbly soils with bare schistose hills rising behind them, and often water below, must lead to unusually intense capture of summer daytime heat and its night-time release to the vines by both convection and direct radiation. Such unique terroirs probably warrant an enhanced soil-type adjustment of at least +1.2°C to the minima. I have not included this among the standard adjustments because of its rarity, and because regionally representative tables in the main seek only to typify regionally good-average viticultural terroirs.

4.6   Proximity to water bodies

Sea and lake breezes have already been partly covered in Section 3.4. The following are supplementary remarks on the immediately local effects of inland lakes and rivers. These gain significance because they can part-offset the limitations of some inland environments: for instance the sustained heat and atmospheric dryness of inland Australia, or winter-spring cold in the cool continental climates of North America and central and eastern Europe.

The role of medium-sized lakes in North America is well established: notably the Finger Lakes in New York State, USA, and Lake Okanagan in British Columbia, Canada. In both cases successful viticulture is confined to the often steep slopes immediately overlooking the lakes. Lake influence in these cases grows by the fact of their elongated shapes, allowing cumulative effects on winds or breezes blowing along them. Temperature inertia of the deep lakes, as of larger water bodies, delays budburst in spring until the worst threat of frost is over, and extends ripening later into autumn. The main effect during ripening is to raise night and minimum temperatures, which according to present argument should benefit grape and wine quality.

European opinion also attributes benefits to sunlight reflected from rivers, such as from the Mosel and Rhine, onto their steep northern banks. Such effects would be especially strong during ripening at or after the equinox, when the sun angle is low. It is a logical supposition for cool, high-latitude environments.

The benefit of proximity to lakes and rivers in hot, summer-arid environments is for the opposite reasons of afternoon cooling and humidifying, although night warming and reduction in spring frost risk also play a part. Even modest-sized water catchment dams can ameliorate such climates slightly for vines sloping down to them. However the biggest benefits come downwind of larger permanent lakes and rivers. As with elongated lakes, breezes blowing along rivers cool and humidify cumulatively. The east bank of Lake Mokoan in north-east Victoria, Australia, is an example of the first type, while in the same region Rutherglen and Wahgunyah get cooling afternoon breezes that blow up the Murray River from the elongated Lake Mulwala. They are marked enough to be known locally as 'The Wahgunyah Doctor'. Similar breezes are thought to be significant for the Nagambie Lakes viticultural sub-region of Victoria's Goulburn Valley, and may also be so upstream and to the east of Lake Eppalock in the Heathcote region.

Benefits such as these are relatively greatest in regions that are well inland as well as being warm and dry. Not only do high evaporation rates maximize air cooling and humidifying within the local convections created, which are themselves strong because of the wide temperature contrasts developed daily between land and water; the convections are also less likely to be overridden by strong regional winds as experienced nearer the coast.

4.7   Topography and soil: summary and discussion

A common thread runs through the topographic and soil features of the known best viticultural sites: all result in greater above-ground temperature equability.

- Slope and good night air drainage raise evening and night temperatures.
- Easterly aspects effectively raise early morning temperatures but reduce those in the afternoons.
- Slopes facing the midday sun are overall warmer than those facing away, but most of this is through soil warming, of which the main above-ground effect is to raise evening and night temperatures.
- Heat-absorptive stony soils reduce above-ground heat load under daytime sun, but increase radiative warming under cloud and during the evening and night.
- Proximity to water bodies cools and humidifies in the afternoon, but warms late at night and in the morning.

Still needing attention are the detailed patterns of change in air temperature and soil-radiated warmth through the 24 hours. I have laid strong emphasis on the timing of afternoon sea breezes as a moderating factor in warm to hot coastal climates, and some on the early warming of eastern aspects. At least equally important could be the patterns of cooling through the evening.

It has been well demonstrated that plant leaves store assimilate during the day that is surplus to what can be immediately used, predominately as insoluble starch (e.g. Fondy and Geiger 1982, 1985). Degradation of the starch to transportable sugars then begins around sunset, so assuring a continuous supply of assimilate throughout the plant for metabolism and growth through the night. The conversion itself requires considerable respiratory energy (Bouma et al. 1995).

Accepting the premise that flavourant synthesis continues through the night, it would seem important that conditions for starch hydrolysis, sugar transport and the biosynthesis and sugar-complexing of flavourants should remain favourable as far into the night as possible. In hot climates this would mean temperatures dropping to the metabolically optimum zone early and staying there through the night. In cool and mild ripening climates the opposite need is for evenings to remain warm as long as possible.

Evening and night conditions gain further significance from the fact that flavourant compounds produced then in the berry skins have least risk of immediate loss through evaporation or heat/light-induced degradation. A continuous night supply of sugar to the skins should also maximize flavourant conjugation into glycosides. Being water-soluble as well as non-volatile, these can then be transported out of the skins to the greater security of the berry interiors.

But for all the circumstantial evidence, we still cannot claim proof that above-ground temperature effects due to topographic and soil features are major contributors to grape and wine quality. The reason is that most are correlated with, and to some extent confounded with, others below ground that are equally key components of terroir. Chapter 5 will examine some of these, chiefly as to how they relate to vine water relations and the development and functioning of the vine root system. Other soil contributions to terroir will be deferred to Chapter 7 (vine nutrition) and Chapter 8, which deals with the wider question of soil types and their relations to underlying geology.

Chapter 5

The Below-ground Environment and Root Function

5.1   Soil water relations

Our understanding of soil water relations as a key component of terroir owes much to the work of Professor Gérard Seguin and colleagues at the University of Bordeaux (Seguin 1983, 1986; van Leeuwen et al. 2004). They found that the soils of superior terroirs have very good internal drainage, whether through coarse texture (sandy or gravelly), or in the case of fine-textured soils, through adequate humus and/or calcium saturation that give a porous crumb structure. Such soils allow abundant, healthy root development throughout the profile and a steady supply of moisture, not too easily extracted, through summer and autumn from subsoil stores built up in winter-spring. Combined with low to moderate fertility levels, this constrains vegetative growth after fruit-set and results in a balanced vine with an open canopy and desirable moderate sunlight exposure for the fruit and new buds. Ability to rely on subsoil moisture in mid to late season helps to insulate the vine and crop from the effects of both summer drought and end-of-season rains, giving greater vintage consistency than on lesser terroirs.

Diverse soil types can provide such a moisture regime, depending on their interactions with prevailing rainfall and evaporation patterns. They include light, gravelly soils allowing very deep but sparse root penetration, in which absorption follows the moisture front progressively down the profile as the season progresses; well-structured heavy soils which have a high water-holding capacity but allow its extraction by roots only with difficulty; duplex soils that perform both functions; and soils developed over chalk or other porous forms of limestone which can absorb and hold substantial amounts of water, and often overlie a water-table. Capillary action slowly supplies water to the limestone surface or to cracks that vine roots can penetrate. Hancock and Price (1990) comprehensively described the role of chalk in this regard.

Recent research in Australia on partial rootzone drying (PRD) has thrown new light on the role of water relations for vine growth and fruiting (Dry et al. 1996, 1999, 2000, 2002; Dry and Loveys 1998; Loveys et al. 1998, 2001). PRD is a technique of alternating drip or similar irrigation on either side of the vine row, which ensures a continuous presence of drying roots on the unwatered sides along with fully watered roots on the sides opposite. Hormonal stress signals from the drying roots, thought to be mediated by abscisic acid (ABA), elicit a partial closure of leaf stomata and a limitation of vegetative growth, particularly that of lateral shoots. This both reduces the vine's water need with little effect on yield, and limits the potential for damaging water stress. Internal transfer of water from the watered to the drying roots appears to be enough to keep the latter viable.

A crucial aspect of the PRD findings is that the roots can only export ABA *during* the process of soil (and hence root) drying. On reaching a critical level of dryness, export ceases. It is not yet clear whether this is because the roots stop producing ABA, or just through lack of water passage to carry it.

From this description it appears that PRD, when effective, essentially mimics the ideal natural water relations that Seguin describes. The latter would likewise produce a continuous supply of ABA from roots under moderate water stress, either from within a soil volume with adequate but tightly held water, or from successively drying soil layers as the moisture front recedes to depth (or both). Section 5.4 carries this argument forward to effects on fruit ripening.

5.2   Soil and root temperatures

Two major recent studies, respectively in the moderately cool Loire Valley of France (Morlat and Asselin 1992; Asselin et al. 1996; Morlat et al. 1997; Barbeau et al. 1998; Brossaud et al. 1999) and the Hawke's Bay region of New Zealand (Tesic et al. 2001a, 2001b) have brought into focus the importance of soil (as distinct from air) temperatures for terroir. In brief, as compared with other soils in the same region and external climate, warmer soils accelerate all stages of phenological development and ripening, and impart superior wine quality. An implied supposition is that this results directly from higher root temperatures and associated soil conditions, not indirectly from radiative warming of the above-ground vine and fruit. Some component of the latter cannot be ruled out, however.

Part of the relationship can, of course, be attributed to better ability to ripen in climates at the cool margin for the grape varieties grown. But that is clearly not a sufficient argument. Were it so, the simple remedy would be to grow earlier-maturing grape varieties or move to a warmer environment, which experience shows is not the case. Other explanations must therefore be sought that are more directly related to root conditions. Four possible mechanisms suggest themselves, probably acting in complementary fashions. The list is almost certainly incomplete.

First, the warmest soils must by definition be well drained, and will usually be stony or rocky or on slopes facing the sun, or all three. Their temperature advantage is most marked in spring, due to rapid warming. This favours early and strong root production of the growth and fruiting hormone cytokinin, resulting in even budbreak, flowering and setting, and hence desirably even veraison and ripening.

Second, such soils, because of their good drainage and usually limited capacity to store water, will reach a state of water deficit and root ABA production by early summer, potentially limiting berry size and lateral branching, and producing an open canopy with good light relations. Barring heavy summer rains, further access to soil water must be by progressive root exploration for deeper or less extractable moisture, which ideally will maintain a moderate stress.

Third, soils that conduct heat readily will normally do so to greatest depth. A well-warmed subsoil will give the most hospitable environment for deep root penetration and dependence on these roots for late-season moisture. This further helps to buffer the vine against drought and unwanted ripening-period rain.

Fourth, the stronger and deeper the warming, the more root metabolism and hormone production will be able to extend into late autumn. This should, as I shall argue below and in Section 5.4, help prolong flavour ripening into times of lower air temperature and hence better retention of the most volatile flavourants.

It is logical to think that strong and deep soil warming will have its biggest impact in the coolest viticultural environments, especially in highly continental climates where winter soil temperatures fall lowest. Table 5.1, from Geiger (1966), shows long-term average soil temperatures at various depths at Potsdam, Germany, in a fairly continental climate just beyond the cold limit for viticulture. The peak temperatures reached annually fall steeply with depth, and average dates of their attainment become progressively later. Those at two metres occur about mid August, and at four metres, late September.

Late-season deep root growth and metabolism must here greatly depend on the intensity and depth of soil warming. This in turn will depend on soil

Table 5.1. Soil temperatures at Potsdam, Germany: averages for 1894–1948.
Data from Geiger (1966).

Soil depth cm	Annual average temp. °C	Average date of highest temp.	Average highest temp. °C
100	10.85	30 July	20.65
200	10.40	15 August	17.20
400	10.00	22 September	13.65
600	9.85	30 October	11.95

thermal characteristics, its surface sun exposure as governed by aspect and slope, and an open vine canopy that lets enough sun through.

We must now ask how far the benefits of soil warming extend into warm and less continental climates. As already noted (Section 4.5), the terroir characteristics associated with maximum soil warming are still sought in warm climates. Even there, soils during early spring growth are often well below optimum temperature for the roots, budbreak having been earlier as temperature allows. Also, many of the warmer viticultural climates have predominantly winter-spring rainfall, potentially leaving the soil wet and cold in spring.

Later in the season, many warm-climate subsoils are still probably cooler than optimum for deep root development and function, upon which many vines in these climates depend critically for their late-season water supply. Stony and rocky soil textures help further by facilitating water movement to these depths and, at the surface, reducing summer water loss by evaporation.

We can again conclude that soils that warm readily, and to greatest depth, are a universal part of good viticultural terroir.

5.3   Development of the vine root system

It is well recognised that root growth substantially controls whole-vine growth and architecture (Richards 1983; Rowe 1993). Evidence already discussed has now shown how it mediates vine responses to water stress via production and export of the stress hormone abscisic acid, ABA.

Rowe (1993) has summarized how the vine root system develops. Roots first exploit the soil avenues of least resistance and largest pores, to establish the beginning of a permanent framework. As these primary roots reach barriers or pores too small and rigid for them to enter, finer laterals develop that can penetrate the smaller pores, followed in turn by still finer tertiaries and so on. Maturity is when the system fully occupies the accessible soil volume.

The mature root system then comprises a framework that grows principally by thickening, with numerous small deciduous roots (feeding roots, or
rootlets) that are in an annual cycle of formation, decay, and new formation.
Available nutrients from decaying rootlets are largely scavenged and in part
stored in permanent vine structures for winter metabolism and renewed
growth in spring. The greater part of the vine's winter-stored carbohydrate
(Bates et al. 2002) and perhaps even more importantly nitrogen (Cheng et
al. 2004) is in the roots. Untapped nutrients in root residues recycle via soil
organic matter.

Growing roots produce the hormones that shape top growth. Gibberellins
(or their precursors) form in the region of cell extension behind root tips
and promote shoot internode growth through cell extension, as well as the
formation of embryonic tendrils rather than fruit clusters in the newly-
forming lateral buds of the shoots (Srinivasan and Mullins 1980, 1981).
Cytokinins, produced within the root tips, by contrast promote new node
and leaf formation, branching of the growing shoots, development of their
existing fruit clusters, and fruitfulness of their newly-forming lateral buds.
Warm soils in spring promote cytokinin dominance and fruitfulness, cool
soils that of gibberellins and vegetative extension. (See Gladstones 1992,
Appendix 1, for a fuller account).

Resulting top growth in many ways mirrors root growth. The young
vine's phase of permanent root extension corresponds with that of rank
shoot growth. Once root growth has passed mainly to the numerous
short feeding roots, cytokinins come to balance the gibberellins from root
extension. Juvenile shoot vigour subsides and the vine passes into its mature
phase of balance between vigour and fruitfulness.

Deep soils with ample water supply prolong framework root development and the phase of juvenile vigour. Shallow soils with limited water-
holding capacity, or soils with texture that obstructs the extension of other
than fine feeding roots, reduce vegetative vigour and advance vine maturity.
The effects of soil bear analogy to those of rootstocks with differing degrees
of vigour promotion.

5.4   The root system and fruit ripening: a hypothesis

Section 5.1 focussed on the influence of the vine root system, as mediated by
root-produced ABA, in limiting vigour under moderate water deficit. This
stems at least in part from ABA's known antagonism to the gibberellins and
cytokinins that variously promote growth. Dominance of ABA also imposes
continued dormancy on the lateral buds of current season's shoots after their

extension stops or slows in mid summer. Later it is responsible for dormancy of the seeds.

Another important role of ABA is now generally thought to be the initiation of grape ripening, directly or through its antagonism to the auxin group of hormones produced by the still-developing seeds (Coombe and Hale 1973 and many later authors). Early research of Weaver (1962) and Hale (1968) showed that applied auxin prolongs the pre-veraison 'lag' phase, or Stage II of berry development. Later Cawthon and Morris (1982) suggested that the known decline of natural seed auxin towards veraison (Coombe 1960) is associated with maturation of the embryos. In this scenario embryo maturation removes an inhibition of veraison that is largely absent from seedless varieties. Length of the lag phase in turn largely governs the earliness or lateness of a variety (Coombe 1976; Mullins et al. 1992). See further discussion in Section 10.3.

Coombe and Hale (1973) found a very rapid increase in berry ABA over the 10 days from the beginning of veraison, followed by a fall over the next 10 days. Later authors (Koussa et al. 1993; Antolin et al. 2006) showed a massive apparent transfer of ABA from leaves to fruit coinciding with veraison. The trigger for this transfer remains uncertain, although declining berry auxin could be one. Rising leaf ABA under moisture stress, or perhaps under declining daylength as shown by Alleweldt and Düring (1972), is another. But while other directly leaf-perceived signals may be involved as well, there is evidence that root moisture stress, where present, is a major contributor to leaf ABA (Loveys 1984; Loveys and Düring 1984).

Whatever the trigger, the effects on the berries at veraison correspond to two of those generally recognised in plants as promoted by ABA: partitioning of assimilate overwhelmingly to the fruit, and later softening of fruit and fruit attachments.

New evidence of Symons et al. (2006) suggests a role here of the brassinosteroids (BR), a recently discovered group of growth substances that is as yet poorly understood. Like ABA these showed a very large increase in berry content at veraison, followed by decline. Application of BR to pre-veraison berries slightly advanced veraison and anthocyanin formation, and increased berry sugar content. A BR inhibitor had the reverse effect. But given what else so far is known of the BRs (Haubrick and Assman 2006), the findings suggest an action most probably confined to a specific promotion of cell expansion during veraison. A more general contribution to ripening seems less likely, since the other known effects of BRs correspond more to those of ripening inhibitors. More research is needed on this point.

We now turn to the later ripening processes, initiated at or after veraison and continuing to fruit maturity.

One incontrovertible fact is that ripening depends on continuing import of a factor or factors from outside the fruit. It stops totally on fruit detachment from the vine at any ripening stage. But while the reasons for stoppage are obvious enough for sugar increase, they are less so for the other components of ripening. These include the biosynthesis and storage of aroma and flavour components (engustment); loss of malic and other acids; continued skin anthocyanin synthesis in red varieties; loss of various herbaceous aromas and flavours; and very importantly, maturation of the tannins in seeds, skins and cluster stems. What imported factors promote these?

Sugar, as the ultimate substrate for synthesis of all other berry compounds, could itself be a limiting factor early in ripening: witness the work of Pirie and Mullins (1976) and others showing close correlations between skin sugar and anthocyanin contents. It also seems reasonable to suppose that adequate sugar is needed early to build up the berry's biochemical apparatus for later ripening.

On the other hand the evidence of Petrie et al. (2000a) showed that although shortage of available sugar can delay veraison, some of the subsequent ripening processes can continue independently of it. These included increases in berry water content and fresh weight, and, contrary to Pirie and Mullins (1976), development of skin anthocyanins.

The evidence seems clearly to counter-indicate sugar supply as primary driver of the later ripening processes responsible for grape and wine flavours. In at least some circumstances (Coombe and McCarthy 1997, 2000) these do not detectably start until after sugar content has peaked. *Current* sugar intake therefore cannot be essential for them. Equally, because flavour ripening stops immediately on detachment from the vine, regardless of berry sugar content already reached, it cannot be driven by existing sugar content. Continuous import of some other factor or factors must be needed.

Two substance classes remain as principal candidates: the hormones (almost certainly), and minerals, the latter as essential components of many enzyme systems of vine and fruit metabolism. For instance the monoterpene cyclases are known to need either magnesium or manganese ions as a co-factor (Gershenzon and Croteau 1993), while iron, copper and manganese play roles in oxidative phenomena and tannin evolution in wine (Cacho et al. 1995), and conceivably could also in the ripening berries.

But whether minerals would have to be in *constant* supply seems doubtful, given that the component minerals for enzyme function, once present, can cycle continuously. The spectrum and relative amounts of available minerals in the fruit could nevertheless still contribute to nuances of flavour, as will be discussed later.

The field therefore narrows to hormones. Their function as rapid-acting chemical messengers demands a sequence of formation, transport and degradation at target sites: an action mode that fits perfectly the observed behaviour of grape ripening and its cessation on fruit detachment. Sources and transport to the fruit could be from leaves or growing points via the phloem system, from roots via the xylem, or from either indirectly via the other.

Which hormones are involved? Auxins from the growing seed embryos certainly inhibit the start of ripening, and it seems possible that after embryo maturity some inhibition might continue from auxins transported down fruit-bearing shoots from still-growing tips. Such growth in any case forms a powerful sink for assimilate, minerals and hormones in competition with the fruit. Auxins may thus retain an anti-ripening function well into ripening. Any significant influence of gibberellins, on the other hand, appears limited to pre-veraison.

Leaving aside the brassinosteroids, as already discussed, the two agents recognised as promoting ripening generally in fruits are ABA and the gas ethylene, C_2H_4 (Sacher 1973). In climacteric fruits (i.e. those, such as bananas, that at ripening exhibit a passing but dramatic increase in respiration and enzyme activity, with rapid hydrolysis of reserve carbohydrate), endogenous ethylene plays a major role. The literature on such fruits nevertheless remains unclear as to whether it is a primary trigger, or whether it forms in response to other ripening triggers and then merely accelerates the process.

In grapes, a non-climacteric fruit, berry ethylene content is low throughout ripening (Coombe and Hale 1973). These authors considered it unlikely to have a dominant role and proposed instead a central function for ABA, at least in ripening initiation. Kanellis and Roubelakis-Angelakis (1993) suggested alternatively that low levels of ethylene may participate synergistically with ABA in the dramatic changes initiated at veraison, but only after ABA had reached a certain threshold. More recently Bellincontro et al. (2006) demonstrated that ethylene treatment of post-harvest grapes can bring about some changes consistent with further ripening. But that does not prove a role in normal ripening on the vine.

One factor which counter-indicates ethylene as a primary agent of normal ripening is that as a gas it can only act on the organs within which it is generated, or at most on others in closed atmospheric proximity. It can hardly carry signals between roots or leaves and fruits. The possibility remains that these could export a water-soluble and non-volatile precursor, which could then generate ethylene within the fruit when triggered by ABA etc. But that should produce a detectable rise in fruit ethylene during ripening, which has not been recorded. The conclusion therefore remains that ethylene can at most have a small and secondary part in normal grape ripening.

Awareness of a crucial function of ABA and its metabolites in fruit ripening generally came well after that of ethylene in climacteric fruits, but has now been demonstrated for several non-climacteric fruits including citrus (Richardson and Cowan 1995), sweet cherries (Kondo and Tomiyama 1998), rambutans (Kondo et al. 2001) and mangosteens (Kondo et al. 2002). But it is still unclear in these instances whether ABA merely triggers ripening, or whether it exercises a more continuing control.

The same question remains for grapes, for which most evidence on ABA still concerns ripening initiation. Nevertheless some evidence does point to roles in the middle and later stages. Ban et al. (2003) found an anthocyanin response to treatment with exogenous ABA at veraison that extended well into ripening. Treatment with auxin, an ABA antagonist, blocked anthocyanin formation over the whole same period. Pan et al. (2005) showed that ABA activates the acid invertases responsible for hydrolysing phloem-delivered sucrose to glucose and fructose in the berry. Similarly Palejwala et al. (1985) showed that ABA can stimulate the conversion of malic acid to sugar in the berry. Downton and Loveys (1978) earlier observed that the greatest fall in titratable acidity and rise in berry contents of arginine and proline (the latter a marker for maturation) coincided with maximum berry ABA.

Further recent evidence has also been consistent with a continuing role of ABA through grape ripening, some showing that it is sourced through the xylem. Anderson et al. (2008) showed an increase in berry phenolics, that was still present at harvest, with application to Cabernet Sauvignon clusters both at and four weeks after veraison. Gu et al. (2008) likewise showed responses with ABA application around veraison, but only if applied directly to the clusters. There was no response from application to adjacent clusters or leaves, suggesting that phloem transport was not involved. Antolin et al. (2008) cite work of Deytieux et al. (2007) showing that from veraison ABA diminishes progressively in the berry flesh but increases in the skin, being present mainly in the skin by harvest. If we accept the skin as being the site of many ripening processes, this again suggests a role of ABA throughout ripening.

While none of this evidence is conclusive, together it givers a fair indication that ABA is the primary hormone imported into grape clusters that both triggers and continues to stimulate ripening.

What might be the source of this ABA? Based on studies with the legume *Lupinus albus*, Pate et al. (1998) posited a typical ABA circulation between leaves and roots in which, under non-stress conditions, main production is in the leaves. But under root moisture stress the root production increases many-fold, to become the primary and dominant ABA source. Research on vines (Loveys et al. 2001) has since shown this to be a response specifically to moderate stress, such as is sustained under PRD.

The pattern fits observed grapevine behaviour. In cool, damp climates at high latitudes, rapidly falling daylengths and temperatures in late summer-early autumn could provide a leaf-mediated brake on vegetative growth and impetus to ripening in the absence of sustained moisture stress. Practical experience and Seguin's Bordeaux research nevertheless suggest that root-zone moisture relations still play a major role there. In warm, low-latitude climates, where all but the latest-maturing varieties ripen under still-high temperatures and at most slowly falling daylengths, control must be more or less entirely by root moisture stress. That agrees with the Australian findings on PRD reviewed in Section 5.1.

ABA is probably not, however, the only root-produced hormone needed for normal fruit ripening. As Stoll et al. (2000) discuss, the hormonal control of vine growth and architecture in mid to late season depends on a balance against cytokinins, also produced in the roots and antagonistic to ABA. Whereas ABA inhibits vegetative growth (especially shoot branching) but promotes deeper root growth, cytokinins promote above-ground growth and enhance resistance to stress, senescence and disease. Maintenance of leaf health and function through ripening is almost certainly a prerequisite for best ripening, as Smart (2006) has argued. As to whether cytokinins have any direct part in the ripening process, no evidence is available at present. But for health of the vine as a whole a balance is required that needs continuing cytokinin function.

We have now reached a point that seems to justify a hypothesis linking root environment and metabolism with grape ripening. It is set out in six parts. The first five encapsulate the arguments already developed; some potential objections are posed and addressed in the following section (5.5). The sixth part relates to nutrition, soils and geology as covered in Chapters 7 and 8.

Hypothesis on root control of ripening

1. All processes of grape ripening are mediated by, or otherwise dependent on, a constant supply of hormones from outside the cluster.

2. The main controlling hormone is abscisic acid (ABA), with a prob-able balancing role of cytokinins to maintain vine and fruit health. Brassinosteroids may contribute to promote berry expansion. Auxins from still-growing seed embryos and shoot tips delay or interfere with ripening.

3. The main continuing source of ABA and cytokinin through ripening is the roots, via the xylem. Leaves, via the phloem, are a complementary source of ABA, mainly for the start of ripening and in cool, high-latitude environments.

4. Optimum hormone balance and intake from the roots depends on continuously mild to moderate root moisture stress, combined with favourable soil temperatures and aeration for root growth and metabo-lism. As will be discussed in Section 6.3, hormone per unit berry weight also depends on the balance between source (root system, especially fine roots) and sink (crop load).

5. Best ripening then depends on favourable above-ground weather: firstly in combination with soil conditions to maintain desired root moisture stress, and secondly above ground for fruit ripening itself. Requirements for the latter vary according to wine style (Table 3.1).

6. As will be discussed in Section 7.3 and subsection 8.3.4, terroir charac-teristics of grapes and wines depend also on the amount and spectrum of trace mineral uptake by the roots active during ripening. Where these are predominantly in the subsoil they convey a consistent terroir imprint reflecting a site's undisturbed subsoil and underlying geology.

5.5   Objections to the hypothesis

The chief objection to a primary root source of ripening hormones comes from work showing apparent blockage of the channels from root to fruit at various stages of ripening. McCarthy and Coombe (1999), Coombe and McCarthy (2000) and Coombe and Iland (2004) suggested that direct xylem connection to the berries becomes fully blocked at or soon after veraison, after which water ingress is only via the phloem along with sugar and other solutes. They further suggested a progressive blockage of the phloem channel, starting about the time of attaining maximum berry sugar, and becoming more or less complete well before maturity in varieties such as Shiraz that are prone to late berry shrivel.

Complete isolation of ripening fruit from the roots has, however, since been disproved. Studies of both water movement (Rogiers et al. 2001, 2006) and dye movement (Bondada et al. 2005) have shown that the xylem-berry connection stays functional to full fruit maturity. Consistent with that,

Rogiers et al. (2000, 2006) showed increase right to maturity in berry content of calcium, which is immobile in the phloem and must therefore reach the fruit through the xylem. Phloem transport does indeed appear to be the main source of berry water and most solutes while berry sugar is still rising; but the evidence of later more or less complete phloem blockage suggests that further accession to the berries must be direct through the xylem: not only of calcium but presumably also of potassium and other minerals.

Having a functioning xylem channel to full fruit maturity makes possible the thesis of late ripening control by root-produced ABA. Supporting evidence can be found in the work of Antolin et al. (2003), who showed a strong relationship in non-irrigated vines between berry and xylem ABA concentrations that persisted to harvest.

The fact that a xylem connection to the fruit, though functional, may have reduced efficiency during late ripening does not necessarily counter-indicate continuing transfer of hormones or minerals. Tanner and Beevers (1990) and Ruhl (1992) showed that transpiration draw plays a minimal part in most mineral uptake and transport within plants. It seems reasonable to suppose that the same applies to root-produced hormones. That is, intake by target organs depends primarily on soil or root supply and their respective sink (i.e. demand) strengths. With reduced water movement into them, hormones and minerals in their transpiration streams must be more concentrated.

Nor does a post-veraison fall in berry ABA, as observed by Coombe and Hale (1973) and others, necessarily argue against a role throughout ripening. High levels are needed at veraison to overcome remaining production of auxins by maturing embryos and/or still-growing shoot tips. Different functions and a relative freedom from antagonists late in ripening could mean that much lower ABA concentrations are still adequate.

Chapter 6

Vine Balances and Management

Chapter 5 developed the hypothesis that grape ripening is regulated by hormones formed predominately in the vine's roots. From that it follows that the extent, health, and physical and chemical environment of the roots must be a major key to best ripening and terroir expression.

But terroir comprises factors of the natural environment both below and above-ground, which must ideally be in balance: with each other, and with the viticulture practised and wine styles aimed at. This chapter first carries forward the arguments of Sections 5.1 and 5.2 to integrate above and below-ground water and temperature relations, then explores the contributions of certain vine and management factors. Chapter 7 considers separately those concerning vine nutrition.

6.1 Soil-atmosphere water balance

Let us accept that the vine needs an ample water supply up to and including fruit set; then a period up to veraison when it ideally has moderate moisture stress to limit berry size (for red wines) and inhibit further shoot growth and branching; and finally, during veraison and ripening, only mild to moderate stress that promotes root ABA production but still maintains vine and fruit health to harvest and beyond. This requires suitable balances at each growth stage of current water supply, canopy size and atmospheric evaporative demand.

Previous discussion underlined the role of well-drained soils with the right degrees of water retention and difficulty of extraction, either throughout the soil profile or in accessible subsoils; or else deep water sources that allow slow upward capillary movement to the roots through limestone etc. There should be enough winter-spring rain (or irrigation) annually to recharge the soil profile and groundwater, then little enough rain in mid to late season to force the vine to depend largely on deep-stored or hard-to-extract water through fruit development and maturation.

But the classic Bordeaux research that led to this conclusion was conducted in a mild and fairly humid environment, with some summer rain. In such climates moisture stresses are seldom damagingly severe even in the warmest and driest summers and autumns, which in general provide the best vintages (G.V. Jones and Davis 2000a, 2000b).

Climates with highly evaporative summer-autumn conditions (high temperatures, low afternoon relative humidities, little cloud) create a different set of problems. Intense vine transpiration depletes soil moisture rapidly. Roots cannot extract tightly bound water fast enough to keep up, nor can they grow fast enough to effectively tap new sources. Resulting root stress, in the absence of irrigation, can then become too severe for the desired production of hormones and/or their export by a failing transpiration stream. Irrigation, if applied, has to be enough to allow unimpeded water uptake to meet the high evaporative demand. Thus ideal moderate moisture stress will be largely confined to fleeting periods between irrigation flushes and intervening strong stresses. If this reasoning is correct, conditions for optimum vine architecture and ripening cannot be met in hot, arid environments whatever the water supply. (Section 6.5 examines this question further, together with the potential of partial rootzone drying (PRD) to circumvent it.)

Conversely, consistently cool, cloudy and rainy weather, or soils that remain wet, must equally preclude adequate root production of ABA. Intermittent occurrence of these conditions during ripening can cause flavour as well as sugar ripening to stall, as is indeed seen in practice.

The ABA hypothesis thus fits practical observation. It also fits the Bordeaux finding of Seguin (1983, 1986) and of van Leeuwen et al. (2004) that in the best terroirs, roots active during grape ripening are deep and sequestered from wetting by all but the heaviest rains at that time. Their ABA production will be less influenced, and can resume more promptly with a return to fine weather and resumption of evaporative demand.

More broadly the ABA hypothesis fits the pattern of ripening-period climatic optima set out by Gladstones (1992, 2004) and here in Table 3.1. For table wines the desired conditions include moderate mean temperatures and sunshine hours, moderately high afternoon relative humidities, and low summer-autumn rainfall combined with least temperature variability. With suitable soils and enough winter-spring rain for spring growth and to replenish subsoil reserves, close to ideal moisture regimes result in most seasons. Exhaustion of soil moisture reserves is balanced in late ripening by falling autumn temperatures and rising relative humidities, provided that the grape varieties are suitably matched in their maturities (see Chapter 10).

An important consideration here is that environmental and, to some extent, management factors can substitute or compensate for each other in promoting desired outcomes. For instance humid afternoon sea breezes, combined with light afternoon cloud, largely compensate for high mean temperatures in allowing the lower Hunter Valley of New South Wales to produce high-quality table wines (ripening-period rains permitting). Similarly some warm-summer, east-coastal regions of the USA can produce good table wines because of their humidity and moderate cloudiness, but only provided diseases can be controlled and sites are best selected to escape winter kill and untimely ripening-period rains.

The most important other issue here is irrigation. Do we accept the hazards of natural rainfall for the sake of closer to ideal cloudiness and humidity, as is the traditional European stance? Or do we go to where the growing season is drier and try to achieve an ideal water supply by controlled supplementary or total irrigation, accepting that these environments may risk excessive atmospheric heat and dryness? Or can terroirs for found that have the best of both worlds? We will take these questions up in Section 6.4 on irrigation strategies. But first we need to examine some other factors that can influence vine roots and their hormonal production.

6.2   Vine and root maturity

Winemakers hold universally that the best grapes and wines come from mature vines. Vine maturity, as we saw in Section 5.3, is effectively reached when the framework of permanent roots reaches its full extent and exploits all the soil volume accessible to it.

An undoubted contributor to the mature vine effect is the expanding capacity with age (to a point) to store reserve assimilate and nitrogen in the trunk and roots. This should result in more even budbreak, flowering and setting, and then, after renewed reserve build-up between setting and veraison, better insurance against over-cropping or unfavourable conditions for photosynthesis during ripening. Having attained its greatest root depth, the vine is also better buffered against wet and drought during ripening.

Reynolds and Wardle (1997) give an apparent example of the contribution of 'old' wood (and presumably root) volume to wine quality. For Riesling vines growing in British Columbia, Canada, they found that increasing volume of old wood resulted in more bound terpenes and increased floral aroma and flavour in both berries and must, together with less vegetal character. This effect was greater than that associated with differences in shoot density and canopy light relations.

The present hypothesis suggests a further relevant relationship. Initial growth of the framework roots of young vines naturally follows the soil channels of least physical resistance and easiest water availability. Thus in the early years of incomplete root development, water supply to the vine tends to alternate between luxury and severe stress. Intermediate periods of optimal stress are short. The opposite obtains for mature vines. Access to the soil's full storage capacity, but with greater difficulty of extraction, extends the periods of optimal stress for root ABA production. So does the mature-vine preponderance of short feeding roots with a high root-tip capacity to form ABA and cytokinins as opposed to gibberellins. The result is a mature spectrum of root-produced hormones associated with vine balance, fruitfulness, and strongest impetus to ripening.

6.3   Vine size and crop load

The argument so far has hinged on maintaining a consistent mild to moderate moisture stress from fruit set to maturity. That applies equally to dry-grown and irrigated vines. Section 6.4 to follow discusses irrigation management. Here we look at the question of vine training and crop load for vines directly dependent on rainfall, which in most viticultural climates imposes a natural limitation on vegetative growth from early or mid summer onwards.

The traditional management in hot and summer-dry Mediterranean environments is one of widely-spaced bush vines, which are pruned hard to maintain small size. A combination of limited leaf area with large available soil volume per vine allows a wide-ranging root system to extract water progressively through the season. Larger vines with more leaf area per unit ground area become possible as growing-season rainfall and/or soil water storage become greater.

Within this context can now be considered the vines' balances of assimilate and ripening hormones. There is no doubt that, under growing conditions that allow it, both leaf area and its effective illumination can be increased substantially by strategies of vine trellising (Smart and Robinson 1991; Smart 1992). That allows commensurate increases in fruit yield while maintaining vine carbohydrate balance and fruit exposure to sunlight. Smart (2002) contends that wine quality can then be maintained to relatively high yields: a position that contradicts traditional winemakers' belief.

Others, such as leading Australian viticulturist and winemaker Brian Croser (personal communication) hold the traditional view to be correct: that quite moderate yield limits exist beyond which wine quality suffers.

Two points from the present analysis tend to support Croser's view.

The first is that, although above-ground canopy can be greatly increased, accessible soil volume and stored water capacity remain constant. Then, just as a highly evaporative atmosphere reduces the chance of continuously moderate root moisture stress, so will the transpirational demand of a large and well-illuminated canopy.

The second point concerns crop load and the vine's carbohydrate economy. A heavy crop load diverts assimilate from root growth. Edson et al. (1995) and Miller et al. (1997) showed, albeit with potted vines, that it can greatly reduce both health and current growth of the roots during ripening. An expected result according to the present hypothesis would be reduced capacity of the roots to produce ripening hormones, especially relative to a large crop to be ripened. The same would apply to trace mineral uptake into the ripening fruit.

Such a source-sink balance of ripening hormones and trace minerals would arguably be highly sensitive to crop load: potentially more so than for assimilate. It would explain how, even though the fruit of over-cropped vines may eventually catch up in sugar content, it fails to attain full flavour.

Imbalance of the vine's carbohydrate economy must be further exacerbated if highly evaporative conditions accelerate sugar transfer to the ripening fruit, as described in Section 3.3. It suggests that one of the problems of grapes ripening under these conditions is not just that sugar accumulates faster than flavour. Competition from too-rapid sugaring, like from over-cropping, may actually suppress the production of root hormones needed for flavour ripening, which as a result cannot begin to gain strength until after sugaring has ceased. This fits the observation of Coombe and McCarthy (1997, 2000).

Bringing these points together, we return to the conclusion that the necessary combination of vine carbohydrate economy, water relations and root hormone production for best wine quality can only be reliably attained where vine vigour is moderate; crop load is moderate for that vigour, so as not to detract from root health and growth; and combined soil and atmospheric water relations conduce to moderate stress post-setting, then consistently mild to moderate stress through veraison and ripening. The ripening period also needs to have low enough evaporative demand for berry sugar accumulation not to outpace other ripening process.

The further question follows: can the desired water and vigour balance be attained using irrigation in dry environments?

6.4   Irrigation strategies

As described in Section 6.3, the ideal of continuous mild to moderate water stress through veraison and ripening is seldom fully attained under natural rainfall. Superior viticultural terroirs give the nearest approach with greatest reliability, but depend on particular soil types and somewhat anomalous combinations of climate elements. Such terroirs are rare and specialized.

Growing winegrapes in dry areas with supplementary or total irrigation has many practical advantages: both in reducing some of viticulture's natural hazards, and in theory by gaining more control over the vine's water economy. In particular, limited supplementary (especially drip) irrigation can make viticulture possible in climates with less risk from rain during flowering and ripening, and on slopes with desirable mesoclimates and freely draining soils that would otherwise be too dry.

European opinion has nevertheless always held that irrigation reduces wine quality. Partly that can be put down to a natural bias of regions whose domestic production does not need irrigation. On the other hand the present analysis does point to valid reasons for the view. We will examine those here and try to identify situations and strategies by which irrigation can be used to best advantage.

In doing so we need to keep in mind that highest quality is not the only legitimate objective. Ability to produce good, as opposed to great, wine in quantity and at costs affordable in the general market for everyday drinking is just as important. Here the cardinal statistic of grapegrowing is the product gained by multiplying yield, quality and the reliability of both, divided by production cost. And terroir evaluation for this equation is as relevant as when seeking terroirs for ultimate quality. Indeed, ideal requirements for both do not differ much, other than in practicability for reliable large-scale production.

The first and overriding point is that climates requiring more than minimal irrigation nearly all have high evaporation rates. Vines then need easy and rapid water uptake if they are to avoid damagingly severe stress, largely by-passing the intermediate stress levels needed for best root hormone production and ripening. Many also have excessive ripening temperatures and sunlight intensity.

The hot, arid inland irrigation areas of California, and to a lesser extent of south-eastern Australia, provide prime examples. Here heat, low rainfall and very low relative humidities dictate a more or less total dependence on irrigation. Water must be enough to maintain a constantly high rate of transpiration. With traditional sprinkler or furrow irrigation techniques this requires frequent replenishment of the upper soil layers over more or less the whole surface area. Roots tend to remain shallow. There is, however, the advantage of some cooling and humidification among the vines.

Drip irrigation under comparable conditions confines root development and functioning to vertical cones beneath the drippers, necessitating perhaps still easier and more constant water availability within them to keep up with transpiration. The same can apply on shallow soils in only moderately dry climates, as Collins et al. (2005) have shown in north-eastern Victoria. It therefore seems clear that no established irrigation system can provide the desired slight to moderate stress, other than transiently, in the typical situations where substantial irrigation is needed.

The question of how far the shortcomings of drip irrigation can be by-passed using partial rootzone drying (PRD, see Section 5.1) or other potential technologies has still to be resolved. Early indications for PRD were promising. But as Gu et al. (2004) pointed out, much of the early research made comparisons only between fully drip-irrigated vines and PRD vines receiving half the water amount; irrigation method was confounded with water quantity.

Later experiments using equal water amounts have given mixed results. Loveys et al. (2001) and Dry et al. (2002) at Adelaide, South Australia, and Antolin et al. (2006) in northern Spain all confirmed positive responses to PRD, while Gu et al. (2004) in the Central Valley of California and Collins et al. (2005) in north-east Victoria both found water amount to be the over-riding factor, with little or no advantage from PRD. Collins et al. showed (on a shallow soil) that roots under drip irrigation were so constricted that the vines passed rapidly in and out of strong stress even with PRD.

These results give grounds for what seems a logical, if still tentative, conclusion. PRD has a useful place, but mainly where evaporative demand is not excessive and the need for supplementary irrigation modest. Its fullest effectiveness also presumably depends on the summer and ripening-period rainfall being low enough not to interfere with the stress regime that PRD imposes.

Seen in overview, a key condition for successfully using drip irrigation of any kind seems to be that there is reliable enough wetting of the full soil profile in winter-spring. The vine needs to develop a fully comprehensive root system under natural rainfall, which taps moisture to depth and from it can supply nearly all its normal water needs for moderate growth and yield. Timely supplements to avoid occasional undue stresses can then be positive for quality without compromising the fundamental architecture of the root system or of the vine's water economy. The advantage becomes clear if this allows grapes to be grown in climates with better surety of growth stoppage before veraison and less risk of ripening-period rainfall. Alternatively a heavy supplementary watering in winter or early spring may successfully mimic a suitable natural rainfall regime, provided that the soil has a sufficient water storage capacity and that the latter part of the season is humid and equable enough not to impose severe evaporative demands.

Dependence on drip or other irrigation to the extent that it starts to supplant natural root distribution must, by contrast, be at the expense of grape and wine quality. And while that sacrifice may still be commercially worthwhile for inexpensive wines, it is incompatible with quality viticulture and especially for individually terroir-based wines. High prices for these depend on their uniqueness and site typicity. As I shall argue in Chapter 8, these characteristics depend on an intimate relationship among roots, subsoil, underlying geology and climate that cannot be reconciled with any major need for irrigation.

Chapter 7

Vine Nutrition

The modern consensus is that climate and soil water relations dominate in shaping viticultural potential. Nutrition is regarded as secondary; it can usually be manipulated and in commercial viticulture is seldom limiting.

Yet important aspects remain in which soil chemistry and to some extent the vine's nutrition cannot be controlled. Deep soil temperature, pH, water-logging and various toxicities are largely beyond the influence of surface treatments, and can limit the effective depth of the vine's root system. This can be critical late in the season, when the vine ideally depends on its deep roots for moisture and arguably for mineral uptake into the ripening fruit (Section 5.5). Whether this is so or not, we have still to explain the many historically established, consistent nuances of wine aroma and flavour that appear largely independent of vineyard and winery management, and to be related instead to soil and underlying geology.

Chapter 8 will look at the broad topic of soil type and geology. Here we will first examine the contribution of nutrition to terroir, starting with the major elements nitrogen and potassium.

7.1   Nitrogen

Little disagreement exists about the role of nitrogen. Soils rich in nitrogen promote too much vine vigour and leafiness. Shading of both fruit and cane bases compromises fruit quality and the fruitfulness of new buds forming for the following season. Vines and fruit are prone to fungal diseases and insect pests (Bavaresco 1989). High nitrogen also increases the ratio of top growth to roots. This accepted wisdom originated in Old World viticulture, where moisture is usually ample and arable soils fairly fertile. Viticulture must therefore be confined to infertile, low-nitrogen soils to achieve vine balance.

These also happen to be mostly stony and sloping, which at the cold limit of viticulture is often essential for grape ripening.

The question becomes more complex in warmer and drier climates, where moisture stresses reduce vigour and increase root relative to top growth. Diverse combinations of nitrogen level and moisture stress can then equally achieve the desired vine vigour and balance, as shown for instance by Choné et al. (2001) in Bordeaux. Choice of grape varieties contributes here: for instance vegetatively vigorous varieties can need less nitrogen and/ or more stress to achieve best balance. Vigour control by rootstocks (Delas et al. 1991; Whiting 2004) gives further options.

Nor is the problem simply one of getting best vine balance and architecture. As has been extensively reviewed (Bisson 1991; Henschke and Jiranek 1991; Kunkee 1991; Rapp and Versini 1991; Sponholz 1991; Bell and Henschke 2005), berries need to accumulate enough nitrogen, mainly in specific amino acids, to support best yeast growth and function during fermentation. Wines made from grapes of adequate nitrogen content generally have superior aroma, flavour and overall quality (Sinton et al. 1978; Goldspink and Frayne 1996; Treeby et al. 1996; Bell and Henschke 2005). And while artificial nitrogen supplements to the must are possible and widely used, there is now good evidence (Goldspink and Frayne 1996; Bell and Henschke 2005) that they fail to enhance fermentation as much as natural berry amino acids.

The amount needed by the yeast also bears relation to the amount of sugar to be fermented, so relative nitrogen deficiency in the must is most likely in warm, sunny climates where sugars are high: especially, perhaps, where good canopy light relations allow maximum photosynthesis and yield from a given leaf area. Some inverse relations between yield and wine quality factors may be traceable to this cause, for instance as found by Sinton et al. (1978).

Berry nitrogen is critical not only for the yeasts. Recent research by Choné (2003) and Peyrot des Gachons et al. (2005) has shown substantial responses to nitrogen application on a low-nitrogen soil in berry contents of the cystcine/thiol conjugates that give rise to key flavourants in the wines of a number of grape varieties (see Section 2.3). This may be due to enhanced availability of cysteine, an amino acid, for conjugation, though other nitrogen requirements for flavourant biosynthesis could also contribute. Equally interesting in this work was the finding that severe moisture stress reduced conjugated thiol content but that moderate stress increased it: a result consistent with the thesis that root-produced abscisic acid modulates ripening processes including flavourant biosynthesis.

An important point for both terroir and its management is that berry nitrogen can also be too high, as Bell and Henschke (2005) review in detail. Excess nitrogen and the resulting yeast vigour and heat output of fermentation can lead to wines with volatile acidity, protein haze, high levels of ethyl carbamate and biogenic amines such as histamines, and an increased risk of subsequent microbial instability. Too-rapid fermentation may also compromise colour extraction for red wines (Treeby et al. 1996).

We are therefore looking conceptually at an optimum of nitrogen concentration in the grapes and must, or perhaps an optimum nitrogen to sugar ratio. This is a further example of how soil and climate can interact in defining the terroirs of grapes and wines.

Given these relationships, timing of nitrogen availability to the vine is critical and has been well studied. That taken up before flowering is used more or less wholly for growth, while uptake between fruit set and veraison or soon after goes principally to the fruit, either directly or after temporary storage in other vine structures (Conradie 1991; Wermelinger 1991; Dukes et al. 1991; Goldspink and Frayne 1996). Then nitrogen taken up after veraison, and wholly after harvest, goes increasingly to storage in the wood and especially roots, whence it is available for winter metabolism and growth the following spring. Such in-vine storage is especially important in cool, continental climates where still-cold soils in spring reduce both the mineralization of soil organic nitrogen and early growth of the roots to absorb it.

The dynamics of nitrogen mobility and timing of root-availability in the soil are exceedingly complex. The problem is simplified in dry climates where drip or like irrigation is practised from fruit set on. 'Fertigation' with urea in the irrigation water can then optimize both amount and timing of summer nitrogen applications, although long-term problems such as soil acidification in the drip zone can arise. Another likely consequence is still greater confinement of roots to the drip zone.

Under full or substantial dependence on natural rainfall this option is not (or not reliably) available. Mostly rainfall can be counted on to wash in spring-applied nitrogen for vegetative growth, but in summer-dry climates that for later berry uptake becomes problematical. Here much will depend on soil characteristics that govern how readily nitrogen is carried down to the root zone and held there. The same factors are important for the breakdown, mineralization and downward transport of nitrogen from winter-spring cover crops or applied composts or manures. These interactions are important aspects of terroir, as are the comparable relationships of other nutrients. We will take them up again in Chapter 8 in relation to soil type.

7.2   Potassium

The relations of potassium supply to vine health, fruit quality and terroir are likewise complex.

On the one hand potassium deficiency can reduce both yield and fruit quality, with uneven ripening, low berry sugar and reduced carotenoid and anthocyanin pigments (Champagnol 1984). Deficiency also exacerbates susceptibility to fungal and bacterial diseases, especially (for some diseases) in the presence of high nitrogen (Huber and Arny 1985; Bavaresco 1989; Marschner 1995). Susceptibility appears to follow from raised contents of cell constituents having low molecular weights, such as amino acids, that pathogenic organisms can directly and readily use. This combines with reduced high molecular weight compounds like proteins and cellulose that give the cells structural strength and organizational integrity. Deficiency reduces the grapes' ability to reach full maturity before disease or senescent breakdown supervenes. Loss of lower leaves in potassium deficiency also leads to fruit sunburn in warm and sunny areas, and hence to further-enhanced disease susceptibility in the event of rain (Greer and La Borde 2006).

Fregoni (1977) reported that vines with ample potassium resist drought well: an observation paralleling that for a number of other crops throughout the literature. But whether deficiency leads specifically to drought sensitivity seems unresolved, although it might logically be expected. The symptoms are similar to those of drought and the two are easily confused. The fact that deficiency symptoms in vines commonly follow summer drying of the potassium-enriched surface soil, leading to late dependence for moisture on the potassium-depleted deeper soil layers (Champagnol 1984), further blurs the distinction.

By contrast the problem in warm, dry areas, and especially with full irrigation, is often that of too much potassium in the berries and musts. It is most acute in red wines because they are fermented on the skins, which contain much of the berries' potassium. High must potassium results in loss of tartaric acid through precipitation as potassium bitartrate, with consequent undesirably raised must and wine pH (Somers 1977; Boulton 1980).

The reasons lie partly in terroir and partly in vine management. Viticultural soils in hot, dry areas are typically unleached and well supplied with potassium. Then, although the evidence is against any major role in potassium uptake of vine transpiration, and therefore of mass water flow to the roots, the soil moisture amply supplied by irrigation does facilitate uptake by the alternative pathway of ion diffusion to the roots (Barber 1968). Supply to the berries is further amplified if flood or sprinkler irrigation encourages root activity to stay in the potassium-enriched soil surface layers throughout the season. Added to this the atmospheric dryness that necessitated irrigation can be expected to accelerate water evaporation from the berries, and therefore translocation to them of potassium from the leaves in the same way as for sugar (Section 3.3).

Two considerations suggest that full irrigation will enhance such translocation still further. The first is that fully irrigated vines tend to develop leafy canopies with many lower leaves shaded, especially if their nitrogen status is high. These will have accumulated potassium while growing, but because of later shading produce little assimilate during fruit ripening. Their solutes unloading into the phloem during ripening might therefore be expected to have a higher than normal ratio of potassium to sugar.

The second is that of sink demand by the berries. Mpelasoka et al. (2003) suggest that under conditions of limited sugar supply these develop a heightened demand for potassium, as the main alternative solute capable of maintaining their turgor and growth. Insufficient sugar supply to the berries is common in heavily cropping irrigated vines with poor canopy light relations, especially if ripening under above-optimum temperatures for assimilation as described in Section 2.2.

In contrast to the lush irrigated vine can be envisaged one in a cooler climate with limited nitrogen and water, with a sparser and well illuminated canopy, and a balanced crop whose sugar demands through ripening (together with enough over for healthy root activity as discussed in Section 6.3) do not exceed available assimilate. Assuming vegetative growth to have stopped, existing leaves will remain fully active throughout (Petrie et al. 2000b). There is no added leaf potassium accumulation by superfluous leaves, nor abnormal demand for it by the fruit. Prolonged and efficient assimilation by the existing leaves means the ratio of total sugar to potassium they supply over the ripening period is maximized.

Mullins et al. (1992) consider in any case that in such 'normal' situations most berry potassium comes directly via xylem from the soil, with little remobilization from other plant parts. That, if correct, has important implications. If root activity by mid-late ripening is largely confined to potassium-depleted deep soil layers, further potassium accession to the fruit will be minimal. Soil differences in this respect must have significance for terroir, as Chapter 8 will discuss further.

7.3   Other nutrient elements
Whereas the contributions of nitrogen and potassium nutrition to terroir are

clear enough, those of other nutrient elements are less so. Gross deficiencies, where identified, are normally corrected. No evidence is available of deficiencies in any of them being beneficial for grape and wine quality. Nor, in most cases, do normally encountered luxury amounts appear harmful.

That said, differences in the supply or balance of nutrients within 'normal' ranges may still influence the subtler nuances of wine character. Given the Old World's many examples of wine flavours apparently related systematically to soil type (see Chapter 8), that seems highly likely. We will therefore now examine some of the literature's better documented findings on plant nutrition that might be relevant. Marschner (1995) is a good general reference.

Most agricultural soils have ample *calcium* for normal growth. Clinical deficiency is rare. On the other hand the evidence indicates important effects over an extended range of tissue contents. Chardonnet and Donèche (1995) and Marschner (1995) note an essential role in stabilizing plant cell membranes, with the susceptibility of tissue to pathogenic attack inversely related to their calcium contents. In storage, fruits high in calcium are more resistant to both physiological breakdown and fungal rotting.

A dominance of calcium among soil cations confers good soil structure and friability (e.g. Maschmedt 2004). Against that, high soil calcium and pH can induce deficiencies of certain trace minerals as will be discussed later in this chapter. Mostly these can be managed without difficulty. The known antagonisms among calcium, magnesium and potassium in plant nutrition also suggests that a calcareous soil or subsoil may reduce potassium uptake into the ripening fruit, which can be an advantage by reducing must and wine pH. In most respects, then, a generous supply of soil calcium is consistent with good viticultural terroir.

Magnesium deficiency can occur on light, sandy soils from which it is easily leached, and can be induced by high calcium and/or potassium. This often happens where light soils receive heavy lime or potash applications. Deficiency causes leaf fall, particularly late in the season. There is also evidence on some soils that magnesium applications or foliar sprays can prevent inflorescence necrosis and late cluster stem necrosis, or shanking (Jordan 1996). Both reduce yield and the latter, fruit quality. On the other hand high soil magnesium contents impair soil structure in the same way as high sodium.

Phosphorus (as phosphate) is fundamental to many plant processes, with deficiency reflected in reduced growth but few specific symptoms. Adequate phosphate is essential for vine health and root development, so that (in contrast to water and nitrogen) nothing is gained by controlling vigour through reduced phosphate supply. Most Australian soils are highly phosphate-deficient for vines as for all crops, particularly soils high in iron, which combines firmly with phosphate and fixes it in plant-unavailable forms. It moves very little down the soil profile except in light sands, so surface applications are generally ineffectual: the more so where the surface soil and roots regularly dry out. Deep banding of substantial amounts beneath the vines as part of initial soil preparation is the best remedy, both to place it where soil moisture and root activity persist longest into the season, and to minimize fixation through its localized high concentration.

Deficiencies of essential *trace minerals* are common in vines. As well as reflecting the compositions of parent rocks, deficiencies can be due either to loss by leaching (mainly from sandy, acid soils) or to various forms of adsorption or chemical fixation by clay minerals that render them unavailable to plants. Soil pH is also a major influence. As illustrated by Robinson (1992) and Maschmedt (2004), copper, zinc, manganese, iron and boron are most readily absorbed by plant roots at neutral to acid soil pH, and diminishingly so as pH rises above neutral. Iron deficiency ('lime chlorosis') is very common in vines on calcareous soils. Conversely molybdenum is most available at neutral and higher pH; while boron again becomes highly available in very alkaline soils, sometimes giving rise to boron toxicity. The high availability of manganese at low pH can give rise to toxicity, as does very commonly that of the non-nutritive element aluminium. Toxicity of aluminium is probably the major cause of poor growth on highly acid soils, and (together with water-logging) particularly of failure by vine roots to survive in deep soil layers. Generally speaking the optimum availability of trace minerals, like that of the major nutrient elements, is within the pH range 5.5–8.0 and coincides with the optimum range for vines overall.

It is likely, if still unproved, that trace minerals play key roles in some of the finer nuances of terroir effects on wine. We have already seen that some are essential co-factors of the enzymes of plant metabolism, including those involved in the synthesis of berry flavourants. Gershenzon and Croteau (1993), for instance, note the need for either magnesium or manganese ions as co-factors of the enzymes that build up monoterpene molecules. The synthesis of such secondary (i.e. metabolically non-essential) products can vary greatly without affecting a plant's fitness, so wide variation in them is possible and normal.

Then again, berry mineral contents and balances are known to affect fermentation, and potentially its flavour products and their subsequent chemical behaviour. Kunkee (1991) noted that adequate must calcium and magnesium are needed to protect against inhibition of fermentation as alcohol content rises. Cacho et al. (1995) showed oxidative processes and the evolution of phenolic compounds in red wine to be influenced by the concentrations of iron, manganese and copper. And as with nitrogen compounds, minerals derived from the berries could well have availabilities and functions in fermentation different from those achievable with artificial must supplements.

Ecological support for a role of trace minerals in defining terroir comes from findings such as those of Greenough et al. (1997) in Canada and Angus et al. (2006) in New Zealand, who showed that the trace mineral compositions of wines could in most cases accurately predict their regional or vineyard provenances. Given the variability of surface soils, such correlations suggest relationships more to underlying geology than to readily observed or measured soil qualities, a point that Pomerol (1989), Wilson (1998) and many others have long maintained.

Chapter 8

Geology and Soil Type

8.1 Introduction

We have so far examined several aspects of soils, emphasizing effects on microclimate above and below ground, effects on root growth and function, and lastly vine nutrition. But we have not yet considered geology or soil type as terroir descriptors in their own right.

A substantial literature exists on the geology of French terroirs. Pomerol (1989), Ribéreau-Gayon (1990) and Wilson (1998) have all published comprehensive and highly informative treatises on the subject. Wilson attempts (with less success) to include climate. Swinchatt and Howell (2004) bring a stronger focus on climate alongside geology in their excellent study of viticultural terroirs in California's Napa Valley. But their conclusions, too, emphasize the big gaps that still exist when trying to integrate climate, soils and geology. Finally I should mention White's text *Soils for Fine Wines* (2003). This brings a soil science perspective as distinct from that of geology. While not addressing terroirs as a whole, it provides an invaluable reference on the soil component.

The importance hitherto attributed to geology and soil type in determining terroir has depended very much on situation. French emphasis on them has been in the context mainly of distinguishing terroirs within areas of subdued land relief and little variation of climate. Terroir distinctions in wine character must then principally derive from soil type and underlying geology, including some control by the latter over local relief and hence mesoclimates. Geology is also a convenient basis for mapping terroirs. As a marketing tool it is particularly useful for implanting images of terroir distinctness.

Associations between wine characteristics and soil type alone, and likewise to some extent geology, have nevertheless proved elusive when studied objectively, e.g. Rankine et al. (1971) and Noble (1979). As already discussed, the best terroirs in Bordeaux include highly disparate soils and geology, having in common only that under the prevailing climate they give closest to optimum water relations. How far these differing soils and geologies can account otherwise for differences in wine style is largely confounded (as in other Old World localities) by the fact that grape varieties tend to be segregated according to the proven best soil types for them.

The fact nevertheless remains that some terroir-related wine characters seem to persist through big seasonal and weather variations. They cannot be related just to water relations or climate.

A study by Wahl (1988) in the cool-marginal climate of Franconia, Germany, attempted to sidestep the problem. By transporting a range of soils to one-cubic metre test containers at nine climatically contrasting test sites, he was able to separate effects of soil from those of weather as recorded at each site. In this case the effects of weather dominated almost completely. Small differences in the wines associated with soil type could be accounted for by differences in soil heating properties and hence fruit maturation.

Such a result still cannot be conclusive, though, because vines grown in natural soils and developing progressively deeper root systems can access terroir factors beyond the reach of those in closed containers. It is in fact part of the present thesis that the deeper roots are critical to terroir expression, which is further consistent with the long-established belief that terroir characteristics in wine only become apparent as the vines mature.

Wahl's study does nevertheless support the general observation that climate is the main terroir factor where climate is most limiting, or where topography provides a locally wide range of climates. It is only where climate differences are small that soils and geology emerge as the dominant component. Further, part of their role stems from interactions with climate in governing root temperatures and moisture relations, as shown in the studies cited in Section 5.2.

8.2 Soils and geology in the literature

The viticultural literature from the early 19th century to the present contains many opinions and observations on soil types or geology, based partly on the science of the time but largely from practical experience. Most that I have been able to access has dealt with French viticulture.

One of the most interesting early accounts is that of the Australian viticultural pioneer James Busby, whose 1825 treatise (based on the French writings of Chaptal and others) and 1833 account of a study tour in Spain and France recorded the essence of French viticultural knowledge at the time. It is worth quoting at some length. The following excerpts are from the 1825 treatise.

> Rich argillaceous [clayey] soils are, in all points of view, improper for the vine; their firmness prevents the dissemination of the minute fibres of the roots, and their coldness is prejudicial to the plant. If a light shower falls, it is evaporated before it sinks beneath the surface, and the same coherence, which prevented the entrance of a lesser degree of moisture, opposes itself to the evaporation of those heavy rains which penetrate deeper. Thus the root seldom receives moisture but in excess …
>
> There are, however, soils rich in nutritious substances, which, containing a larger proportion of siliceous or calcareous matter, do not partake of the hurtful qualities of those in which argil predominates. In these, the vine grows freely, but this very strength of vegetation … is essentially hurtful to the quality of the grape, which, with difficulty, attains to maturity, and produces a wine without strength or flavour …
>
> Calcareous soils are in general favourable to the vine; dry, free, and light, they every where afford a free circulation to the water with which they are impregnated, and allow the numerous tender ramifications of the roots to extend in all directions … Their stimulating nature, too, which, while it increases the energy of the plant, does not impart to it an excess of nutriment, points them out as peculiarly fitted for its culture. Accordingly, we find the soils of many celebrated vineyards to be calcareous …
>
> A mixture of stones is always an important addition even to a soil possessing all the requisites of dryness, lightness and porosity. The root spreads itself easily in a soil rendered penetrable by a mixture of rounded stones, and while the bed of pebbles on the surface oppose themselves to the evaporation of the necessary moisture, they facilitate the filtration of what may be in excess, and reflect back on the grape, the benign influence of the sun's rays.
>
> Volcanic soils have been always observed to produce delicious wines. These *virgin* soils having long elaborated in the interior of the earth by subterranean fires, present an intimate mixture of all the earthy principles. When this semi-vitrified substance is decomposed by the combined action of air and water, it furnishes all the elements of a good vegetation, and the fire with which it is impregnated, seems to pass successively into all the plants confided to it …
>
> There are many places … where granite has ceased to retain that character of hardness which generally distinguishes that primitive rock; and where, pulverized by the action of the elements during many ages, it is reduced to a sand of finer or coarser description. The soil of the vineyards of Hermitage,

and of many others of great celebrity, consists of this decomposed gravel, which seems to possess all the requisites for a superior produce of wine ...

It may be concluded ... that the vine may be cultivated advantageously in a great variety of soils ... The sandy soil will, in general, produce a delicate wine, the calcareous soil a spirituous wine, the decomposed granite a brisk wine.

French writers of the mid to late 19th century, such as Rendu (1857), Petit-Lafitte (1868) and Portes and Ruyssen (1886) placed particular emphasis on the benefits of stony soils, mainly for their influence on vine temperatures. Thus Petit-Lafitte wrote [my translation]:

According to M. de Gasparin these stones, because of their intimate composition, have the advantage of conserving the heat given them by the sun for long after sunset ... thus giving the plant a less sudden transition from day to night and adding to the daily duration of temperatures favourable to its fruit.

Petit-Lafitte cited many additional regions besides Bordeaux, hot as well as cold, where stoniness was known to promote wine quality, including the 'bituminous schists' that enable ripening as far north as the Belgian province of Liège. Both Rendu (1857) and Portes and Ruyssen (1886) developed similar themes, and included detailed physical analyses of the soils of leading French viticultural regions. Portes and Ruyssen noted as well the favourable effect of iron oxide on clay subsoils. "It is on beds of red clay that lie many of the best vine soils."

Fregoni (1977) gave a comprehensive review of earlier terroir studies and conclusions. Mostly they confirmed what can now be taken as established: that vine water relations are paramount, and that, at least in cool climates, stoniness and soil colour contribute importantly to vine microclimate above and below ground. Beyond these he noted a number of points of seeming consensus.

- The wines from stony soils have both high quality and high alcohol content.
- Wines from calcareous soils are high in alcohol and strongly perfumed.
- Clay soils produce wines high in extract and prone to coarseness.
- Sands produce light, aromatic wines.
- Wines from humus soils are coarse, have little aroma and keep poorly.

Like Busby, Fregoni canvassed an apparent relationship between wine quality and soils or subsoils rich in iron and other minerals; and further, that

aromas can differ systematically with the geological formations from which
the soils or subsoils were formed. This seems consistent with a role of trace
or other minerals in flavourant synthesis, and also with Busby's predilection
for volcanic soils. Interestingly, though, Fregoni dismissed the common belief
that red soils best suit red wines, and pale or grey soils, white wines.

One case that seemingly supports that belief is found in the Hunter Valley
of New South Wales, where upwards of a century's experience has shown
the red variety Shiraz (syn. Syrah) to have particular adaptation to slopes or
outcrops of red soil on basalt. The white Semillon, by contrast, gives its best
wines on the alluvial pale duplex, but somewhat gravelly, soils of the valley
floor (Lake 1964; Halliday 2006).

Several leading modern French researchers have commented on matches
between soils and wines. Mostly they agree with those of earlier French writers.

Peynaud and Ribéreau-Gayon (1971) wrote as follows.

> On rich soils adequately supplied with water, the grape-vine yields heavy
> crops but the quality of wines is poor. Most of the best vineyards are generally
> encountered on poor and stony soils. Chemical analysis of soils of the most
> famous wine-growing regions in France indicate that the levels of inorganic
> nutrients are surprisingly low. However, since the grape-vine is able to send its
> roots into the soil to depths of 12–24 feet, the amounts of inorganic nutrients
> available to the roots may be considerable. If the humidity of the soil at the
> root level is high during the maturation period, the grapes are richer in acids,
> particularly tartaric acid, than if soil humidity is low. Clay soils produce more
> acidic, less delicate grapes, rich in pigments and tannins. Limestone soils give a
> particularly high content of odoriferous constituents to some varieties of grapes.

Huglin (1986) reviewed a number of previous publications offering
contradictory views, and concluded that vine-soil water relations dominated
in shaping terroirs, together with thermal factors in the coolest regions. He
considered the evidence for relationships to soil mineral composition to be
weak. Huglin did, however, note that in Burgundy the best Pinot Noir wines
come from limestone soils, the best Chardonnay wines from more clayey
soils derived from marl, and the best Gamay wines (in Beaujolais) from
granite-derived soils. Moreover of two main soil types recognized in Haute-
Bourgogne, having similar elevations and exposures, those derived from
limestone with stony surfaces typically give highly perfumed wines, whereas
those with a finer-textured surface give wines that are more tannic and
less rounded, and with a different bouquet. There appeared to be a good
correlation between soil clay content and astringency in the wine.

Champagnol (1984) wrote extensively on the expressions of terroir, and while acknowledging the role of water relations, still placed considerable emphasis on soil mineral composition and the individual reactions of grape varieties to it. Thus he held Pinot Noir, Chardonnay, Sauvignon Blanc and Cabernet Franc to attain their fullest character on calcareous soils, but Cabernet Sauvignon, Chenin Blanc, Gamay, Grenache and Carignan their greatest fullness and finesse on acid soils. (The case of Chenin Blanc was presumably based on southern French experience.) He also pointed out that whereas calcareous soils are highly regarded in northern France, in the south it is the red soils derived from schist, sandstone or conglomerates that give the more harmonious wines, despite their water relations being supposedly inferior to those over limestone. The key in the south, he believes, is redness of the soil, indicating oxidation of their iron content and hence good drainage. He further suggested (as have many other French authors) that true terroir differences in wine only appear at the highest levels of quality, and that they are probably related to soil mineral composition.

Galet (2000) made many comments on soil. Some of the more relevant are as follows:

> Sandy soils are good for white wines, since they tend to provide finesse and full bouquet. However, good red wines can also be produced on sandy soils (Sables-Saint-Emilion) ... and even give rise to *grand cru* vineyards if some pebbles are mixed in ...(e.g. Graves).
>
> High-clay soils retain water for long periods during the winter, whereas in summer they dry out extensively. The resulting cracks in the soil are harmful to the root system. Soils high in clay are cold, difficult to plough and are suitable only for the production of ordinary wines.
>
> Limestone soils are generally low in organic matter, since lime activates nitrification ... The presence of limestone in vineyard soils is favourable for quality wine production. Many famous wines come from regions with limestone soils: Chablis (Kimmeridgian limestone marls), Vouvray (Turonian micaceous chalk) and Saumur (mix of Senonian, Turonian and Cenomanian). Similarly in the Champagne region, the best wines are obtained from grapes harvested on the most calcareous parts of the Senonian. It seems that aroma development is activated by the presence of limestone in the soil ... The geologist Coquand (1858) showed that the quality of Cognac spirits is directly related to the limestone content of soil in the Grande Champagne vineyards of Cognac (Santonian and Campanian).

Gravelly soils are not very fertile, since there is little fine earth. They are relatively unevolved, skeleton-like soils that originate from the alluvial deposits of a river or sea. As a result, high-quality grapes can be obtained, though to the detriment of yield ... The effect is most noticeable when the soil surface is composed of large stones ... Such is the case in some vineyards at Chateauneuf-du-Pape ... Attempts are made to bring the large stones to the surface.

Maximum temperatures, which are highly related to reflection of sunlight, are higher above light-coloured soils ... (They) can become excessive, and the resulting day-night temperature differences are more frequently harmful above light-coloured soils than above dark ones. Light-coloured soils are suitable for white varieties, but the planting of black varieties on such soils is usually avoided, since black grapes heat up more quickly than white ones and are thus more prone to sunburn.

Finally can be mentioned some experimental observations in South Africa, as reported by Saayman and Kleynhans (1978) and van Zyl and van Huyssteen (1984). In a strongly winter-rainfall climate with little or no irrigation, they found that the soils giving best water relations and most consistent wine quality parameters were of duplex type. Of the contrasting red or yellow soils with a more uniform profile, those with stones gave lower yields but better quality. A key difference was that the clay subsoils of the duplex soils were able to conserve moisture and make it slowly available during ripening, whereas vines on the uniform-textured soils used it up too quickly and then suffered strong stresses during ripening.

8.3   Some conclusions and discussion
In this section I shall try to draw together the salient evidence and opinions on soils to see what generalizations are possible.

8.3.1   Soil structure and drainage
All evidence confirms that both surface and internal soil drainage must be good for vines to succeed. It should be combined with water-holding properties in the deep root zone that allow adequate storage of water and its steady, slow release to vine roots through fruit development and ripening.

Sandy or very stony soils in warm and hot climates must be deep so as to have enough water-holding capacity. Again there should be a retentive subsoil, or else a deep water table or seepage, the retreat of which through the season allows steady limited uptake by sparse deep roots and moderate stress in remaining roots. It is essential that the deep water not be stagnant, nor the subsoil too acid or that it contain highly available toxic elements. In very cool climates, on the other hand, shallow and calcareous or rocky soils with limited water-holding capacity are more appropriate.

For heavy soils with a high water-holding capacity, but also as a rule more difficulty of water extraction, the important factor is physical structure and friability to allow free drainage and root permeation. Comprehensive root permeation in turn allows progressive extraction of all water that is root-available. In contrast are soils with a 'blocky' clay structure, that allow easy root penetration and water extraction from cracks, followed by failure of the roots to penetrate the remainder. Coarse cracking of clay soils as they dry also injures the roots. Cass et al. (1998), Cass (2005) and Cass and Roberts (2006) give a particularly useful treatment of these aspects. Importantly, the latter two papers extend the concepts to the uptake of nutrients as well as water.

The best medium to heavy soils and subsoils are those high in calcium and/or iron, both of which help promote good physical structure. Surface soils will have enough, but not excessive, organic matter (see Chapter 9) to encourage earthworms and other soil-opening microfauna, together with microflora that produce cementing agents for a stable soil crumb structure. Stones and rock fragments will keep them open, assist water infiltration, and protect against surface erosion and evaporative water loss. Soils with most or all of these properties allow permanently sustainable viticulture. Clearly excluded are any soils too high in sodium (sodic) or magnesium (magnesic), both of which destroy structure and upset nutritional balance, or that have toxic levels of elements such as nickel, aluminium (at low pH) or boron (at high pH).

Intermediate duplex or gradational soils, i.e. those with clear or progressive changes to heavier-textured subsoils, can share all the virtues of the contrasting types described above, provided always that they are well drained and that they allow a depth of root development and moisture access appropriate to the needs imposed by climate. In summer-dry climates a low surface water-holding capacity may be beneficial because surplus winter-spring rainfall passes through to a more retentive subsoil, where it is protected against surface evaporative loss and is slowly available to deep roots later when most needed.

8.3.2 Soil thermal properties
As already discussed in Sections 4.5 and 5.2 there can be little doubt that soils that readily absorb heat, and either re-radiate it at night and under cloud or

transmit it to depth, are highly desirable for viticulture in most (perhaps all) climates.

The need is most critical at the cool limit of viticulture, where shallow, rocky soils allow the greatest heat transfer to a relatively shallow root zone, and also limit water storage and availability. Included here are the shallow, chalk-based soils of the northern mid to high latitudes, which, because of chalk's porous nature and very free drainage, enjoy both good subsoil warming and an optimally slow supply of water back to the deepest roots by capillary movement. Hancock and Price (1990) cover the issues in detail.

A related point of interest is the ancient vineyard practice in Champagne, described by Wilson (1998), of spreading 'cendres noires', the 'black ashes' of locally occurring soft lignite. Wilson emphasizes the mineral contribution these make, but equally important could be that their dark colour increases heat absorption. Naturally dark-coloured rock-based and rocky soils that drain and warm readily achieve the same, for instance the slate soils of the Mosel Valley in Germany and the schistose soils of Central Otago in southern New Zealand.

We can see now why shallow calcareous soils are not valued in warm to hot climates. Surface-exposed limestone reflects back too much heat to the fruit, at least for red grapes. With mostly hot and dry summers, any advantage of early rootzone warming in shallow soils is likely to be more than offset by lack of enough deep water storage to meet summer and ripening-period needs. Nor are the limestones of these climates of the same favourable type as in Northern Europe, not being crumbly through fracturing by ice-age freezing and thawing. Although calcium saturation remains beneficial for soil and especially subsoil structure, much the same is assured by the high iron content of many of the best warm-climate soils.

Good soil heat absorption is still important in warm, and even hot, climates. Except under full or near-full irrigation, roots that remain functional there during ripening are often at considerable depth, where summer warming barely reaches and temperatures can still be below optimum for root metabolism during ripening. Good absorption also reduces daytime heat load on the fruit, and through later re-radiation helps to maintain favourable above-ground warmth for ripening through the night.

8.3.3   Soil types and vine nutrition

The relationship of soil type to vine nutrition changes with season and vine growth phase. Ideal nutrition for vegetative growth differs from that during ripening and after.

Accepting that early spring growth depends entirely on vine reserves from the previous season, assimilate needs from then on must be met by the maturing shoots and those for soil nutrients by new feeder roots that start to develop only after budbreak. At first this will be mainly in or near the soil surface, since that warms first and is best aerated. It is also the richest part of the profile for most nutrients, whether from natural re-cycling or from surface fertilizer application. The degree of surface concentration will be most for the least soluble nutrients such as phosphorus. Potassium is intermediate, concentrating fairly strongly at the surface of soils high in clay colloids and organic matter, but leaching readily in sands. Inorganic nitrogen is the most mobile. In root-available form it leaches quickly from the surface of all but heavy-textured soils, unless first captured by vine or other roots or where drought halts all leaching.

The availability of soil organic nitrogen depends on breakdown of the more nitrogen-rich forms of organic matter (e.g. from recent plant and root growth, especially that of nitrogen-fixing legumes) to form ammonium ions (ammonification). Nitrification then oxidizes ammonium to nitrate, which is highly mobile and open to both leaching and plant uptake. Both steps depend on the activity of soil bacteria. White (2003) gives a concise account of these processes.

Dependence on bacterial activity has direct implications for terroir. Low soil temperature, low pH and poor oxygen supply all inhibit. Combination of some or all of these factors often leads to spring nitrogen deficiencies in regions with cold winters: that is, until the soil warms and dries enough in late spring. It is one reason for the cool-climate preference for shallow chalk-derived and similar soils, that favour nitrification with their guaranteed drainage, neutral to alkaline pH and early spring warming.

These factors have less relevance in warm climates, where (in the absence of spring waterlogging) nitrogen supply can sometimes become too great and result in unbalanced growth as discussed in Section 7.2. This is to varying degrees avoided where the surface soil is sandy or gravelly and subject to leaching, or where growth is checked early enough by moisture stress. Non-leguminous spring cover crops can often achieve the same result by competing for both nitrogen and water.

All these implications for terroir tend to reverse with the summer change-over to advanced fruit development and ripening. Generally, and especially in warm and sunny climates where berries reach high sugar and potassium concentrations, the need then is for more nitrogen and less potassium. Desirably, active root function is by then largely in the deeper soil layers.

We can see here a further possible reason why duplex and gradational soils are well suited to viticulture in warm to hot, winter-spring rainfall, summer-dry climates. Moderately light-textured topsoils, for instance a gravelly sandy loam, can have enough binding capacity to retain nearly all potassium in the surface layer where it is available as needed for growth and early fruit development. Roots active later, after the surface soil has dried out, are in the potassium-depleted subsoil, which minimizes further uptake when the ripening fruit has become the main nutrient sink. At the same time winter - spring leaching of nitrogen from the surface layer may be enough to limit vegetative growth, while its capture by the heavier-textured subsoil provides a reserve for uptake late in the season when it goes beneficially to the fruit, or afterwards into vine reserves for the following spring.

The generally very insoluble trace minerals may slowly concentrate at the surface, but remain mostly adsorbed to or as part of the subsoil clay and underlying parent rock (if present). Such rock is usually rich in trace minerals, especially if of volcanic origin. Vines growing on these soils will then have an ample supply of trace minerals during ripening when their roots are active mainly at depth.

8.3.4 Relationships to geology

Geological processes shape the landscape and provide the raw materials from which soils form under the influences of climate, vegetation and time. The end product of these processes, combined with current climate, shapes favourability or otherwise for viticulture.

But this still does not explain the subtle differences of wine character that are consistently reported to be associated with underlying geology. Even allowing for commercial myth-making, the evidence and opinions reviewed in this chapter do very largely support their existence.

Logically the reason for their consistency must lie in factors that change little with vineyard management (fertilizer use etc.) or with season. This brings into focus the nature of the subsoil, perhaps especially the transitional zone of subsoil formation from underlying rock, which would be at the limit of root penetration and largely reflect the composition of the parent rock. Trace element uptake from this zone, late in the season when the ripening fruit is the vine's main sink for nutrients, would seem to provide the most plausible basis for soil-related terroir character. This conclusion is, of course, far from new. I suggest it requires three suppositions.

1. That particular wine characters are associated with specific amounts and balances of root-available trace minerals.
2. That their uptake during fruit ripening has particular significance.
3. That this is when phloem transfer from other plant parts is closing down and mineral nutrients enter the fruit principally from the deeper soil and roots via the xylem.

As well as being consistent with reported terroir relationships to underlying geology, such a mechanism further corroborates two traditional observations: firstly, that wine terroir distinctions only become fully evident as the vines mature and their root systems reach their fullest extents, and then that during ripening they have to depend largely on their deepest roots for moisture. These water relations in turn produce most ripening hormones and highest wine quality. Terroir expression and quality do go together.

We can also see why soils in which either the whole profile or subsoil are rich in nutrient minerals, such as were formed recently over volcanic or similar rocks, produce wines full of character and 'fire' as the early writers describe. The iron and perhaps other basic trace minerals they have in abundance conceivably play a special part in the formation and/or metabolism of phenolics in red grapes and wines (see Cacho et al. 1995, as noted in Sections 5.4 and 7.4). Their normally good structure and drainage allows deep and comprehensive root exploration. At the same time their usually red or dark surface colour benefits red grapes in particular by absorbing daytime heat rather than reflecting it immediately back to the clusters. All these points are enhanced where the subsoil or whole soil profile contain much rock or gravel; these not only increase soil heat absorption, but also mean that the vine roots are in constant contact with a trace element source often governed by local geology. All points support the common belief that such soils are specially suited for red wines, and that they promote a strong sense of terroir in their wines.

Sandy soils do on the whole appear to produce more delicate wines than heavy soils, and therefore to be suitable for white rather than red wines. But given good drainage the vine roots may still colonize mineral-rich deep layers in time, as in Bordeaux; or they may reach ground water that can often be rich in iron and other minerals.

These are at best only broad indications. Many other factors enter the equation. Individual grape varieties have specific needs: for instance the

shallow-rooted Semillon, on its own roots, must have good moisture availability at not too great a depth. Naturally vigorous varieties need soils of low
fertility to reach the vine balance that gives best quality; weaker-growing
varieties reach the same balance only on more fertile soils. Then if mineral
nutrition is indeed a major basis for terroir differences in wine character,
grape varieties can be expected to differ in their responses according to the
synthetic pathways of their distinctive flavourants. All these factors can to
some degree be modified using rootstocks and other management practices,
and interact with climate above ground.

Chapter 9

Organic and Biodynamic Viticulture

9.1   Organic viticulture

9.1.1   *Definitions*

First we have to resolve what the term means in common usage, which is contradictory and confusing.

'Organic' viticulture, as it has been formally defined, authorizes only traditional forms of disease, pest and weed control, using products deemed to be natural and eschewing all that have been artificially synthesized. The contradiction lies in the fact that chemically the chief sprays and dusts allowed (copper compounds and sulphur, for fungal disease control) are inorganic, whereas the prohibited synthetic compounds are all or principally organic, i.e. carbon-containing. Nor is distinction possible on the ground of toxicity. For instance toxic build-up of copper in the soil has long been reckoned a problem in Bordeaux, while sulphur applications can be toxic to useful microfauna and can hasten soil acidification.

The distinction is clearer for soil management and plant varieties. Organic viticulture aims to feed and build up the soil, and thence, indirectly, the vine. It does this with composts, mulches, and inter-row green manuring crops or plant cover. Foliar feeding using natural products such as fish emulsions is also allowed. It prohibits artificial fertilizers and plants developed using genetic engineering technology.

Other vineyard operations such as trellising, pruning, irrigation, harvesting and physical weed control are as in standard commercial viticulture. For concise but fuller accounts see Allan (1998); Waldin (2006a). Note that the present discussion does not encompass organic winemaking, which is an unrelated issue.

9.1.2   *Some theoretical advantages*

To be available for uptake by plant roots, mineral nutrients have to be present in soluble inorganic forms or some very simple organic complexes, or else in forms that can break down to these under the influence of root excretions or microbiological action within the root-hair zone (rhizosphere).

Readily soluble inorganic fertilizers, especially in sandy soils, are available for quick uptake but are subject to leaching loss that can pollute dams, streams and groundwaters. Most trace minerals, on the other hand, and phosphate in iron-rich soils, quickly form physical and chemical bonds to the finest (colloidal) clay particles and become immobile and largely unavailable to plants. They tend to remain either at or close to the soil surface, and so out of reach to active roots when the surface dries out, or, in the case of trace elements, still largely as part of underlying rock or subsoils forming from it.

As a result it is hard with inorganic fertilizers to maintain a balanced, continuous availability of essential nutrients by normal soil application. Application with drip irrigation (fertigation) is possible where irrigation is practised, but this defeats the aim of encouraging a deep and wide-ranging root system.

Suitably chosen organic sources of nutrients can largely obviate these problems. First, because the materials are directly or ultimately plant-derived, the balance of nutrients they supply will bear some relation to the needs for new plant growth. Second, availability of the nutrients depends on organic matter breakdown, the rate of which depends on much the same factors as govern potential rate of plant growth, i.e. soil temperatures and aeration in the region of the root system. Nutrient availability therefore matches the vine's needs, resulting in a steady plane of nutrition and best chance to manage for a balanced vine.

The implications extend further. As reviewed by Bavaresco (1989), plants weakened by clinical or marginal nutrient deficiencies, or by nutrient imbalances such as excess nitrogen, succumb more readily to many diseases and pests. A transient deficiency or imbalance can provide a 'window of opportunity' for these, after which control becomes more difficult. A continuously balanced plane of nutrition optimizes the vine's natural defences to ward off incipient attacks.

9.1.3   *Demonstrated advantages*

Some of the advantages of organic soil management are well known and long established.

A soil's organic matter content contributes to its friability, permeability to air, water and root growth, and water-holding capacity. In soils of low clay or calcium content it can be the primary source of these qualities.

Also well established, but not widely enough appreciated, is the role of organic matter in feeding and maintaining worm populations, and of the latter in opening up channels for air and water infiltration and in re-distributing organic matter and nutrients through the soil profile. Surface application of composts and readily biodegradable mulches stimulates underlying worm multiplication, and transport by them of organic materials deeper into the soil. Worms also effect downwards incorporation of surface-applied lime or gypsum, an action that itself makes the soil more hospitable to them (Buckerfield and Webster 2001a).

Organic matter at appropriate stages of decomposition is essential for a flourishing soil microflora, which plays many roles in vine and other plant health. Among these are production of chelating agents and organic acids that render iron and other largely insoluble trace elements available for plant uptake (Hoitink et al. 2002). Some rhizosphere micro-organisms produce hormones that directly stimulate plant growth. A rich and varied soil microflora also provides natural competitors and antagonists to root pathogens, while some evoke systemic resistance to leaf as well as root disease (Hoitink et al. 2002; Weckert 2002).

A flourishing soil microflora (and microfauna) helps to break down the residues from annually renewed vine roots, together with the short-lived roots of other plant cover. This adds to the work of earthworms in opening up channels for infiltration and new vine root growth.

Field trials and observations in Australia (Buckerfield and Webster 2000, 2001b, 2002) and New Zealand (Mundy and Agnew 2002) have demonstrated major advantages to vines from the application of composts and readily decomposed mulches, with effects from single applications lasting up to several years. But straw mulches, that do not quickly decompose, have proved more problematical, particularly in cool climates where they can retard soil warming, nitrogen availability and vine growth in spring (Ludvigsen 1995). It is also important that all organic applications be made with discretion. Too much can easily lead to excess available nitrogen, to the detriment of vine balance and of fruit and wine quality (Morlat 2008; Morlat and Chaussod 2008; Morlat and Symoneaux 2008).

Does organic soil management enhance terroir expression? The reasonable answer is that it probably does, if we accept that ripening-period mineral uptake from decomposing parent rock, and from the deeper soil layers most closely resembling it, plays a part. According to the above arguments organic soil management should allow the deepest and strongest root penetration into these layers and extraction of minerals from them. Other factors contributing to this are reduction of soil compaction by wheeled machinery, and water supply or management that forces the vine to rely on deep moisture reserves during ripening.

9.1.4 *Integrated Pest Management*

Integrated Pest Management (IPM) is the name given to a developing field of knowledge and practice that should quickly become part of mainstream viticulture. Unlike formalized organic viticulture it does not absolutely prohibit specified types of chemical treatment. Rather, it seeks to mobilize and encourage natural predators of vineyard pests, and, where supplementary chemical control is still needed, to ensure that it does least harm to the predators (and humans, and beneficial soil fauna) but enough harm to the pests for practical control. In this way the use of toxic pesticides can be greatly reduced or even eliminated.

To do this requires that suitable plant hosts for beneficial microfauna exist close to the vines. IPM is probably most effective where summer rainfall is enough to maintain companion plants among the vines throughout the season. The alternative in summer-dry climates is to maintain permanently vegetated corridors and headlands through and around the vineyard.

Diversity of the companion plant population is important to maintain a broad spectrum of beneficial fauna, that can respond to the greatest variety of pest species. However the most appropriate botanical compositions will vary with their adaptation to the environment, their hospitability to desired microfauna, and the nature of the dominant pests to be countered. Achieving the best balances for individual cases will call for careful research and management.

Bernard et al. (2007) give a useful overview of IPM and set out cumulative steps towards its implementation. In brief summary these are as follows.

Step 1: Minimize sprays that are toxic to beneficial insects, spiders and mites. Monitor pests and diseases.

Step 2: Monitor beneficials to guide spraying needs. If possible, confine to spot spraying.

Step 3: Encourage beneficials by growing under-vine, inter-row or corridor sources of nectar, pollen and shelter.

Step 4: If needed, raise and release supplementary beneficials.

Step 5: Encourage wider landscape vegetation to harbour background

populations of beneficials, many of which can travel considerable distances. Maintain local diversity of native flora, and in time
perhaps enrich it with superior hosts for beneficials.

To these steps can be added the introduction of macrofauna such as
Guinea fowl or domestic fowls, which have been used with some success to
control larger insects in the vineyard.

9.1.5 *Practicability*
The wide adoption of organic viticulture will depend on its practicability. Its
philosophy needs to be as already developed for integrated pest management,
rather than in having rigid (and partly illogical) rules for formal certification as at present. This applies as much to the proscription of 'genetically
modified' plants as it does to chemicals used in spraying. A limited market
among committed believers may continue for wines produced and certified
under the old rules at likely premium cost; but far more important is that
the more sensible parts of organic practice become entrenched in mainstream
viticulture.

The aim is ecologically benign vineyard management within a viable
business framework. This requires more skilled management than previous
systems with heavy chemical reliance; but it is repaid in the longer term by
almost certain improvements in wine quality, terroir expression and vineyard
sustainability.

Some compromise is usually needed. An example is that generally the
soils most suited to quality viticulture have low, or at most only moderate,
fertility. Some otherwise very suitable soils are severely deficient in key plant
nutrients such as phosphate and certain trace elements. It makes good sense
as far as possible to remedy these deficiencies early, using whatever fertilizers
are most effective. This allows not only healthy growth of the vines, but also
that of the green manure crops etc. upon which a subsequent fully organic
regime must depend.

Bruer (2001) makes the valid point that inorganic methods with generous
fertilizer use and chemical weed control can as a rule achieve faster and safer
vineyard establishment than organic methods. The resulting time difference
can be economically decisive. He advocates using inorganic methods up to
near vine maturity, then converting to organic.

9.2 **Biodynamic viticulture**
Biodynamic practices encompass all those of organic viticulture as formally

defined, but add a philosophy and certain measures that purport (although the connections are obscure) to grow from it. Koepf et al. (1976) give a comprehensive and generally balanced account covering all agriculture. Joly (1999) deals specifically, if emotively, with his pioneering experience of biodynamic viticulture in the Loire Valley, France. For a concise account see also Waldin (2006b).

The biodynamic movement grew from a course of lectures by the Austrian philosopher Rudolph Steiner (1861–1925) to a group of Silesian farmers in June 1924. Previously Steiner had edited the scientific writings of Johann Wolfgang von Goethe (1749–1832), whose holistic approach to plant biology became an important influence on his thinking.

Biodynamics sees the farm as an integrated and self-sufficient biological unit. Its components – crops, pastures, animals, soil and above and below-ground natural flora and fauna – are ideally in harmonious equilibrium. Vines in such context will express terroir to a maximum degree, since all the major inputs of composts, mulches and animal manures come from the same land.

Biodynamics also emphasize the spiritual connection between the farmer and his land, and, more controversially, cosmic connections with the wider universe. Regarding the first, Koepf et al. (1976, p. 29) state: "The person who actually works on the land has the advantage of direct experience of the more intimate interplay that exists in the growth of plants, the rhythms of the seasons, the thriving of livestock, and in general the communities of the field or garden. A personal relationship to all these arises out of his daily work." Implied is a deep respect for all natural forms and interrelations of life, both within the farm and more widely.

The controversial, indeed notorious, part of biodynamics concerns the practices that it claims influence the reception of 'cosmic' and 'earthly' forces that control biological processes, including plant growth and reproduction. In the case of viticulture the practices consist partly of timing vineyard and winery operations according to phases of the moon and its placement relative to signs of the zodiac. The other part lies in the use of preparations that are variously sprayed, in minute quantities, on the vines or the soil or in compost making.

The best known of the preparations, called no. 500, consists of cow manure that has been packed into cows' horns (according to Joly, never bulls' horns) at the autumn equinox and buried over winter. The horns are dug up after the spring equinox, the contents of one being enough to treat one hectare of soil. It has first to be 'dynamized' by stirring in luke-warm water, preferably by hand and over a precisely selected hour: first in one direction to create a deep vortex, then in the other. The chaotic movement of the water on change in stirring direction supposedly allows the influence of cosmic and earthly rhythms to be imprinted in it. Thus dynamized, the liquid is strained and sprayed on to the soil surface between 3 p.m. and sundown, most effectively when the moon is in front of an earth/root sign such as Taurus, Virgo or Capricorn (Waldin 2006b). The main claimed effect on viticulture is to stimulate soil microbial life and humus formation, and to encourage deeper vine rooting.

Preparation no. 501 is said to work in a contrary way. Here the horn is packed with finely ground quartz and buried at the spring equinox, for conditioning over summer. A hectare needs only part of a teaspoonful. After digging up at the autumn equinox and dynamizing as above, it is sprayed directly on to the vines early after sunrise. Contrary to the 'earthly' effects of preparation 500, this is said to enhance cosmic influences, with stronger photosynthesis, more upright vine growth, and fuller ripening of the fruit and wood.

The final preparations, numbers 502–507, are based on various plant materials such as camomile blossoms, oak bark and stinging nettles, and are used as activating agents in compost making. It is claimed that compost so treated makes the soil more receptive to lunar, solar, stellar and other rhythms.

Joly (1999, p. 66) also describes a biodynamic method for controlling insect and vertebrate pests.

> Incinerating an insect or the skin of a vertebrate on a precise date makes that type of life form flee at the end of the third year. In reality, they can no longer reproduce in this location … The way of proceeding with these incinerations is very precise: twenty or thirty insects, which can be dead, are put into a wood fire. When the fire burns out, the ashes are saved and dynamized for an hour to obtain a uniform mixture. The product can be spread in either pure or diluted form …

Joly notes further, without going into detail, that the date of incineration is critical and should respect a 'planetary' calendar, and that the type of wood used for incineration may play a part because is also has a planetary influence. Similarly spreading of dynamized incinerated seeds can stop plant types from reproducing. However, while stating that the treatments are effective, he acknowledges that their mechanisms remain unexplained.

Objective evidence that any of these treatments work is lacking. Anecdotal reports of positive responses in viticulture have typically followed conversion to biodynamics from earlier systems depending heavily on chemical use, the harmful effects of which Joly so forcefully (and justifiably) denounces. Such responses are to be expected, as they would be from conversion to simple organic viticulture. The real question is: do biodynamic treatments improve on the latter?

Reeve et al. (2005) describe a trial in California to answer that question for one set of conditions. Planted in 1994 and with treatments delineated in 1996 in four randomized blocks of 0.6 hectare plots, the trial compares a standard organic regime with the same plus biodynamic treatments. The paper reports measurements made in 2001, 2002, and 2003.

No differences could be found in any measure of soil physical, chemical or biological qualities, including those of soil microbial metabolism. The biodynamic plots yielded 39% more earthworms, but owing to extreme sample variability this was not statistically significant. No differences could be discerned in leaf nutrient compositions or in cluster or berry weights.

Yield comparisons were obscured by the fact that, as part of a commercial planting, all plots were crop-thinned to a target yield of 10 tonnes per hectare. Some small differences nevertheless showed up in a measured final yields and vine balances. The organic treatment averaged slightly higher yields and lower pruning weights, suggesting greater fruitfulness. Against that, its yield/prunings ratios overall were marginally in the over-cropping range, whereas those for the biodynamic treatment achieved the range considered ideal for quality. Consistent with this difference, the biodynamic fruit averaged slightly higher contents of sugar and phenolics on the common harvest dates, but as with the yield data the differences attained at most marginal statistical significance. The authors concluded that: "Based on the fruit composition data, there is little evidence that biodynamic preparations contribute to grape quality."

The objective case for biodynamics in the vineyard thus remains 'unproven', with little to indicate difference from an equivalent organic regime. Certainly the theoretical grounds for biodynamic treatments, based as they are in pre-scientific metaphysics and astrology, can be dismissed as nonsense. If results do differ it will be for other reasons. Nevertheless an attraction to biodynamics will undoubtedly survive among those, producers and consumers alike, who are drawn to the occult or are antipathetic to science.

It cannot be denied that biodynamic wines can be, and many of those now available are, excellent wines. Nothing in the biodynamic treatments is likely to do harm, while the rituals of their use may, as in religion, inspire to a heightened commitment. Spiritual satisfaction for true believers may itself justify higher costs of production and essential certification for a limited market.

The question of biodynamics in mainstream viticulture raises wider issues. As I have argued in Section 9.1, adoption of organic methods as far as practicable, including integrated pest management, presents a largely proven pathway to better wine quality and terroir expression, and, most importantly, to improved vineyard sustainability and ecological status. No real evidence exists that using biodynamic preparations adds to this. To the contrary, in the cynical world of modern business the chief risk is that a superficial (or claimed) adherence to biodynamics will be merely a spurious selling point, to the potential detriment or neglect of genuinely sound viticultural practice.

There is a final and deeper reason why biodynamics should be questioned. While parts of Steiner's philosophy are admirable in their original context, what has come down to the present is a set of mediaeval superstitions that science has long superseded. At best they are a harmless conceit, not to be taken seriously. At worst, they represent an unhealthy retreat into irrationality and mysticism, such as the world has too much suffered from in the past. They have no valid place in an enlightened 21st century.

Chapter 10

Maturity Rankings of Grape Varieties

10.1 Introduction and historical background

Matching grape varieties to terroirs requires, as a first step, a knowledge of their relative maturities. Only then is it possible to estimate their phenological fit to known or estimated climates. Other adaptive features do, of course, contribute to their performance in given climates; but the overriding consideration in a preliminary evaluation must be when (or whether) they ripen on average, and therefore the expected average conditions under which they ripen.

The literature on grape maturity rankings presents many problems. Some stem from differences in the climates and seasons in which reported observations have been made; others from differences in winemaking purpose, and hence definitions of grape maturity; others still, from confusions over varietal identity.

Some writers, e.g. Jackson (1993b) and Due (1995), have questioned whether consistent maturity rankings exist. Due even argues (Due et al. 1993; Due 1994) that vine phenology is unrelated to climate, and must be established uniquely for each combination of variety, terroir and management. Both objections obviously call into question the value of studies such as described here and in Gladstones (1992). The issues have been extensively debated in the Australasian literature (Jackson 1993a, 1993b, 1995a, 1995b, 1996; Due 1994, 1995; Gladstones 1994, 1996). I show there that Jackson's and Due's views were wrong and based on misapplied statistics.

French researchers of the 19th and early 20th centuries accepted that grape varieties could be classified according to their relative maturities, and that maturity rankings were largely consistent across environments. Rendu (1857) wrote of varieties having early, medium or late maturity, though only in terms of typical maturity dates in their respective normal environments, without allowing for differences across environments. Puillat (1888), and later Boursiquot et al. (1995), avoided that problem by making all comparisons in the one warm environment of Montpellier, southern France, where all varieties ripened successfully. Puillat classed varieties into four 'epochs' of ripening, corresponding to early, midseason, late and very late. He also noted subdivisions within epochs for some varieties.

Puillat's classification and terminology have since been generally accepted. Importantly, experience across diverse environments has largely confirmed his rankings and their universality.

Some phenological interactions of varieties with other environmental elements besides temperature, and with management methods, do undoubtedly exist; but the practical record shows that compared with that of temperature their combined effects are small. Their existence does, however, require that definitions and varietal comparisons of maturity should as far as possible be under comparable conditions of vine husbandry and commercial purpose.

The two assumptions made here are: first, that the vines are sited, pruned, cropped and (where necessary) watered for moderate yields consistent with high-quality winemaking; and second, that maturity is defined as that appropriate for making dry table wines. Early picking for sparkling wine or for rosé wines from red varieties, or late picking for sweet dessert and fortified wines, can be factored as desired around the maturities and maturity dates for dry table wines. Section 11.8 gives more details.

10.2 Revision of the maturity groupings

In *Viticulture and Environment* (Gladstones 1992) I defined eight maturity groups, requiring E°days ranging from 1050 to 1400 at 50 E°day intervals. With a maximum daily accumulation of nine E°days, i.e. from effective daily means of 19°C or over, this meant that in warm to hot climates the standard interval between group mid points is 5.6 days and the total span from earliest to latest group mid points just under 40 days.

Experience since in using the scale has suggested that it was too compressed. Support for that conclusion comes from the work of Boursiquot et al. (1995), who recorded maturity dates for 2168 *Vitis vinifera* varieties in the INRA collection at Montpellier. Ripening temperatures there are above 19°C for all varieties. Discounting a few extreme outliers, the seasonally averaged span of maturities was about 60 days.

The revised maturity groupings set out here (Table 10.1) comprise nine groups at 60 E°day intervals, requiring from 1020 E°days for the earliest Group 1 to 1500 for Group 9. With ripening at effective 19°C or above this

Table 10.1. Grape variety maturity groups and E°days needed for their maturation.[1]

Group and required E°days	White grapes	Red grapes
1 1020	Bouvier, Madeleine Angevine, Madeleine-Sylvaner, Optima, Ortega, Reichensteiner, Schönburger, Siegerrebe	
2 1080	Auxerrois, Bacchus, Breidecker, Chasselas, Faber, Huxelrebe, Müller-Thurgau, Muscat Ottonel, Seyval Blanc	Blue Portuguese, Meunier, Zweigelt
3 1140	Albariño, Chardonnay, Ehrenfelser, Elbling, Frühroter Veltliner, Gewürztraminer (Savagnin Rosé), Kerner, Morio-Muskat, Muscat à Petits Grains (Muscat de Frontignan), Ondenc, Pedro Ximenez, Pinot Blanc, Pinot Gris, Sauvignon Blanc, Scheurebe, Silvaner, Sultana, Verdelho	Bastardo (Trousseau), Chambourcin, Dornfelder, Gamay, Pinot Noir, Tinta Amarella
4 1200	Aligoté, Fernão Pires, Feteasca, Grüner Veltliner, Leanyka, Melon, Muscadelle, Riesling[2], Rotgipfler, Sauvignonasse (Tocai Friulano), Semillon	Aleatico, Dolcetto, Durif, Grolleau, Lemberger (Blaufränkisch), Malbec, Pinotage, Tempranillo (Tinto Roriz), Tinta Amarella, Trollinger (Schiava), Zinfandel (Primitivo)
5 1260	Arneis, Chenin Blanc, Furmint, Marsanne, Rkatsiteli, Taminga, Tulillah, Viognier, Welschriesling (Italian Riesling etc.), Zierfandler	Cabernet Franc, Cinsaut, Lagrein, Merlot, Shiraz (Syrah)
6 1320	Colombard, Crouchen, Folle Blanche, Hárslevelü, Macabeo, Malvasia Bianca, Palomino (Listan), Roussanne	Alvarelhão, Barbera, Cabernet Sauvignon, Corvina, Kadarka, Mondeuse (Refosco), Ruby Cabernet, Sangiovese, Tannat, Touriga Nacional, Valdiguié
7 1380	Airén, Garganega, Gros and Petit Manseng, Mauzac, Trebbiano (Ugni Blanc)	Carmenère, Croatina, Graciano (Morrastel), Grenache, Grignolino, Mission (Criolla, Pais), Mourisco Tinto, Nebbiolo, Petit Verdot, Ramisco, Sciacarello, Sousão, Tyrian
8 1440	Bourboulenc, Clairette, Grenache Blanc, Muscat of Alexandria (Muscat Gordo Blanco), Picpoul, Rabigato, Sercial, Terret Blanc and Gris, Verdicchio	Aglianico, Aramon, Canaiolo, Carignan, Montepulciano, Mourvèdre (Mataro), Plavac Mali, Raboso
9 1500	Biancone, Bombino Bianco, Doradillo	Negroamaro, Tarrango

[1] Biologically effective degree days to make dry table wines: see Section 2.1.
[2] At its cool limit, for classic riesling styles. Elsewhere, probably Group 5.

corresponds to an interval between group mid points of 6.7 days, and a total span between earliest and latest groups of 54 days: i.e. in fairly close agreement with Boursiquot et al.

In allocating varieties to maturity groups I have used information from a wide variety of published sources. More important sources include Rendu (1857), Puillat (1888), Marès (1890), Viala and Vermoral (1901–1904), Galet (1958, 1962, 1979), Branas (1974), Winkler et al. (1974), Norton (1979), Robinson (1986, 2006), Kerridge et al. (1987–1988), Coombe (1988), Dry and Gregory (1988), Mannini et al. (1997), Ministry of Agriculture, Fisheries and Food, France (1997), and Jackson and Schuster (2001). The last reference is especially useful on recently bred varieties for cool climates. Many passing references and hints exist in the general literature, while towards the cool limit of viticulture it is possible to draw fairly detailed inferences from the locations and mesoclimates in which varieties have been commercially successful.

An advantage of having nine maturity groups rather than eight is that they adapt better to common terminology: early, midseason and late, with three subdivisions within each. Thus within midseason, Group 4 equates to early midseason, Group 5 to average midseason and Group 6 to late midseason, and so on. There is in fact little change from the previous groupings for individual varieties, the main differences being expansion of the group intervals from 50 to 60 E°days, and the addition of a very late group containing only a few varieties. Because most of my information comes from English and French-language sources, some important eastern European varieties are regrettably not included.

10.3   Physiology of maturity differences

Section 5.4 has already touched on the hormonal processes leading to and perhaps controlling ripening. Our focus now is on those that result in varietal differences in ripening time. As noted earlier, maturity rankings here refer to varietal groupings in their relative times of maturity for making dry table wines. This is an arbitrary end-point for a succession of phenological stages through which the vine passes each season.

Quite large differences already exist among varieties in their timing of budbreak, and these in part carry through to contribute to maturity differences: see for instance Coombe (1988) and Pierce and Coombe (2004). And because the timing of budbreak is governed both by temperature-controlled dormancy through winter and its release by rising temperature in spring, it follows that some interactions will occur with patterns of winter-spring temperature change: i.e. between continental and maritime climates.

Resulting errors in prediction of average maturity dates from this source nevertheless appear to be small, given reasonable uniformity of pruning time, water relations and crop level as already specified. The main basis for varietal maturity differences still lies in the interval between flowering and veraison, which is under direct temperature control.

Winkler (1962) and Coombe (1973, 1976) recognized three stages of berry development from the time of setting to maturity. Stage 1 is of rapid early berry growth, spanning the last cell divisions and following cell growth in the pericarp (flesh plus skin) plus early seed growth. Auxins and gibberellins produced by the seed embryos, once established, seem adequate to attract nutrients for further pericarp development. This accounts for the correlation between seed number and final berry size (Coombe 1960; Ebadi et al. 1996). In the case of seedless berries, Coombe concluded that such growth as they make probably depends on gibberellins imported from outside the berry.

In Stage 2 there is little further growth of the pericarp, the berries remaining small, hard and opaque green. This is known as the lag phase. Consolidation of the seeds and growth of the embryos continues, as does their auxin production.

Stage 3 starts at veraison, which marks the beginning of ripening. There is a sudden berry softening, followed by a dramatic expansion and water intake, and change from opaque green to translucent light green. Hexose sugar accumulation begins, and a week or so later the first skin anthocyanins appear in red varieties.

The literature is divided on whether embryo maturation necessarily precedes veraison. Hale (1968) and Cawthon and Morris (1982) consider that it normally does; indeed, Hale suggested that the post-veraison stage is essentially one of senescence. Conversely Peynaud and Ribéreau-Gayon (1971) and Mullins et al. (1992) emphasize the variability of seed maturity relative to berry ripening, stating that in late varieties the seeds can be a fully mature and hardened well before fruit ripeness, whereas in some early varieties they are still not viable at harvest.

In any case the known decline in berry auxin content at veraison (Coombe 1960) and the finding that applied auxin delays ripening (Weaver 1962; Hale 1968) suggest that the main barrier to ripening during Stage 2 is auxin produced by the growing embryos. The early ripening of seedless grape varieties, and of seedless berries of seeded varieties, supports this conclusion.

Other evidence underlines the major complementary role of the 'stress' hormone abscisic acid (ABA) in triggering ripening (Coombe 1973, 1976;

Coombe and Hale 1973). Section 5.4 has already discussed this in some detail, and in particular a postulated continuing role of root-produced ABA in promoting later ripening processes. ABA forms in plants under a variety of stress conditions, including heat and drought, and in response to declining day length. It is antagonistic to the growth-promoting auxins, and when sufficient causes a slowing and ultimately cessation of growth and a diversion of assimilates and other nutrients to storage tissues in the fruits. The process is analogous to the 'self-destruct' mechanism of annual plants as they mature and die.

In summary, we can envisage declining auxin production by the maturing seed embryos, balanced against ABA production at least partly related to environmental stress or other signals. Tipping the balance in favour of ABA precipitates veraison, which irreversibly establishes the clusters as the vine's dominant sink for assimilates and probably minerals, and initiates ripening.

Varietal differences in length of the individual berry growth stages are surprisingly little documented, beyond recordings of flowering, veraison and maturity dates and some general observations.

Studies on the length of Stage 1 of berry development are particularly lacking. Varietal differences are in any case typically obscured by low spring and early summer temperatures which ensure that growth rates to the end of Stage 1 are largely temperature-controlled.

It appears generally assumed that, under comparable conditions, the length of Stage 3 (ripening) is fairly standard among varieties: but that this is likewise obscured by factors such as varying crop loads and falling temperatures. Also, nearly all past measurements have been of sugar accumulation, the rate of which depends not only on temperature and crop load but also on atmospheric humidity (see Sections 3.3, 6.1).

More interesting would be the interval between veraison and flavour (i.e. non-sugar) maturity. Few, if any, unequivocal measurements of this have been reported. A provisional impression is that late-maturing varieties do on average ripen more slowly, and need higher night temperatures to do so, than early-maturing varieties.

What seems generally agreed is that the main basis for varietal maturity differences still lies in the interval from flowering to veraison (Coombe 1988; Pierce and Coombe (2004); and within that, particularly the length of Stage 2 of berry development (Mullins et al. 1992; Kanellis and Roubelakis-Angelakis 1993). The view finds clearest illustration in the early ripening of seedless varieties, and within varieties, of seedless versus seeded berries.

The big maturity differences among seeded varieties have still to be accounted for. Presumably the embryos in seeded varieties develop and mature at differing rates, or else there are genetic differences in seed number and/or auxin production. Alternatively varieties may differ in their production of, or responsiveness to, balancing ABA in triggering veraison. These matters await elucidation.

How accurate, then, can the varietal rankings and their E°day requirements in Table 10.1 be, given likely interactions in budbreak time and other potential sources of error? Some varieties will in any case be borderline or only fall marginally within a ranking. Nevertheless the record, at least for quality viticulture, still shows generally good consistency across environments for the rankings as tabulated. Those for the better-documented varieties should probably be accurate to within a maturity group either way; those for relatively unknown varieties, perhaps to within two maturity groups either way.

10.4 Maturity rankings and grape phenolics

10.4.1 *Berry tannins*

The term tannin covers a range of polyphenolic compounds, of which the basic monomeric subunits are flavan-3-ols. These build up by serial addition of extension subunits to form procyanidins, or condensed tannins, in the skins and seeds: a process that is most active early in berry and seed development (Kennedy et al. 2000; Downey et al. 2003a; Robinson and Walker 2006).

According to these authors, and based mainly on studies with cv. Shiraz, total extractable tannins build up to reach a maximum per berry a little before veraison in the skins and a little after in the seeds (mainly in the seed coats). After that the extractable amount falls progressively in both skins and seeds, but most steeply in the seeds, to only about half of the original by maturity. This is thought not to represent an absolute loss, but rather a modification by oxidation or binding to other cell constituents.

Average length of the polymers differs markedly between skins and seeds. That in the skins has normally ranges around 30 subunits. Seed tannin polymers are much shorter, mostly between three and six subunits. Views differ on changes in polymer length with maturation. Downey et al. (2003a) suggest increasing length early but some shortening between veraison and harvest. By contrast Kennedy et al. (2002) found increasing length throughout maturation, at least partly by incorporation of anthocyanin molecules. Unlike those in the skins, seed tannins retain a significant proportion of monomer catechins to maturity.

Gawel (1998) has discussed in some detail the question of tannin astringency and bitterness. He concluded that while bitterness is clearly a taste sensation, that of astringency is tactile only, a feeling of roughness, or 'grip' on the palate that results from interaction with salivary proteins and loss of their lubricating function. Kennedy et al. (2006) have since demonstrated a close correlation between the perceived astringency of tannins and their ability to precipitate protein.

The relations of polymer length to astringency and bitterness are complex, but as a broad generalization it appears that astringency increases with polymer length, while bitterness is associated more with the monomers and short polymers of the seeds.

Thus tannin 'maturation' during berry ripening might be seen as partly due to falling extractability of all tannins, but principally to the steep fall in extractability of those contained in the seeds that contribute most to bitterness. The latter process is signalled by readily perceptible changes in seed colour and texture (Kennedy et al. 2000). It may also parallel to some extent the maturation and lignification of the vine canes, all being promoted by ABA.

What, you may ask, has all this to do with vine maturity rankings? The potential connections are several, but three stand out.

First, the period of tannin synthesis ends approximately with veraison, of which the timing is largely determined by the length of Stage 2 of berry development. Late-maturing varieties therefore have longer for tannins to build up. Further, there is evidence (Kennedy et al. 2000) that the build-up in seeds switches during the latter part of Stage 2 from building polymeric forms to accumulating monomers: a strategy which has doubtless evolved to assure maximum bitterness and unattractiveness to predators through early ripening. Perhaps the long Stage 2 of late-maturing varieties allows this to happen to greatest degree, whereas in early varieties the accumulation is foreshortened and reduced: hence their notable vulnerability to predators. On both grounds the final tannic bitterness of late-maturing varieties is potentially greater.

The same line of thought gives rise to a further speculation. If the timing of veraison is additionally influenced by factors such as crop load, water relations and strength of the ABA signal, it follows that whatever brings about early veraison for a given variety might also result in 'sweeter' tannins. Early veraison then also allows a longer period for reduction in seed tannin extractability and for other components of ripening. Might this be one reason why early vintages are generally good vintages?

Whether intrinsic varietal differences exist in their rates of physiological or flavour ripening remains largely unexplored. As already noted there is some reason to think they do, with early varieties ripening faster and able to do so at lower temperatures. This may be related to a weaker physical structure with lower tannin content, and/or less resistance to ABA from seed-produced auxins.

Any such differences would have important implications for wine style. Rapid engustment, from varietal as from environmental causes, notionally increases the surplus of volatile flavourant synthesis rate over that of loss by evaporation or degradation. This favours accumulation of all volatiles, but especially those with the lowest boiling points and the most readily degraded. By contrast slow ripening allows the retention only of more stable flavourants with high boiling points, although over time they may build up to considerable intensity.

Various combinations of these factors can explain the observed broad relationship between grape maturity ranking and potential wine style. For white wines, winemakers as a rule seek the least possible tannins but maximum grape-derived aromatics, which together come most naturally from early and quick-maturing varieties in environments giving cool late ripening. To retain their low boiling point aromatics they must also be fermented cool. The aromas can then be perceived in wines drunk cold. Most (but not all) late-maturing white varieties, suited only to hot environments, produce wines lacking in aroma.

Paradoxically a few early-maturing varieties, such as Verdelho, can produce quite aromatic wines in hot climates. This might perhaps be explained by particularly rapid flavour ripening, or else superior ability to convert their aromatics into non-volatile, water-soluble conjugates that can move to the berry interior and are thus protected. Restrained yields and a plentiful supply of conjugating sugar or cysteine molecules would assist this.

For red wines the requirements are different. Red wines owe their essential characteristics to phenolics and to flavourants or their conjugates present largely in the skins, the latter being necessarily resistant to evaporation and therefore of high boiling point and stability. This is especially so since coloured berries heat more than white ones.

Because polymeric tannins form only before veraison, that period must be long enough to accumulate the required amount. In most early red varieties it is insufficient to make other than light, low-tannin wines. Pinot Noir is the only one widely used for serious red wines, and it achieves adequate tannins only in very specific terroirs.

Red wines from late-maturing varieties, with a long berry Stage 2, are generally more tannic, although reduced tannin extractability due to prolonged ripening may offset this. Aromas are subdued and only become apparent at relatively high drinking temperatures (as would historically be the case in their natural areas of production).

The best varieties for mainstream red wines, then, are of intermediate maturity, mostly from maturity groups 4 to 7 as shown in Table 10.1. They represent a range of compromises and styles. Earlier ones such as Merlot, Malbec, Tempranillo and Cinsaut tend to produce soft wines that can mature quickly and depend on their primary fruit qualities or roles in blends with more austerely tannic wines. Later high-quality varieties such as Cabernet Sauvignon, Petit Verdot and Nebbiolo are more tannic; some are commonly blended with earlier-maturing varieties to hasten maturation and softening. However exceptions exist, such as the relatively early but tannic Durif and the fairly late but commonly low-tannin Grenache; while factors such as yield, heat and water stress all influence the eventual tannic outcome.

10.4.2 Anthocyanin co-factors

It has for some time been recognized that the colour stability, density and hue of red wines depend not only on concentration of anthocyanins, but also on their aggregation into evolving polymeric or oligomeric structures with other phenolic wine constituents (Somers 1971, 1998). This is called co-pigmentation. It results in more intense and richer colours than can be accounted for by anthocyanins alone. The aggregation is initially in a loose equilibrium with free forms, and immediately reversible with heating or dilution; but in time more stable forms evolve that maintain wine colour long after simple anthocyanins have disappeared. The importance of phenolic co-factors is evidenced by the very close correlations found in well-made red wines among colour intensity, quality ratings, and total phenolics as measured by their UV absorption profiles (Somers 1998).

Boulton (2001) has comprehensively reviewed knowledge of co-pigmentation in grapes and wine to that time. Rather than anthocyanins forming associations with tannins or phenolics generally, as many had previously thought, he concluded that just two types of molecule are significantly involved. These comprise firstly the monomeric catechins, or flavan-3-ols, which are the building blocks of the polymeric tannins but also persist free in quantity in the seeds (see subsection 10.4.1 above); and secondly the flavonols such as quercetin, which are also monomeric and occur only in the berry skins. Both types of molecule have basic structures similar to that of the anthocyanins but are non-pigmented or (in the case of flavonols) at most light yellow. Importantly their concentrations in musts and wines are nearly always much less than that of the anthocyanins, so that the potential for co-pigmentation is limited directly by their supply. Also, because of their bitterness in free form, it can reasonably be suggested that progressive incorporation with anthocyanins into stable larger molecular structures contributes to wine palate softening and reduced bitterness. (Remaining bitterness would then be related primarily to the short tannin polymers and monomers ex the seeds.)

The skin flavonols are of particular interest. Because they bind much more strongly with anthocyanins than the catechins, they are the more effective co-factors. But also their concentrations vary enormously with terroir and vine management, together probably with varietal maturity ranking as I shall argue below.

Price et al. (1995, 1996) and Haselgrove et al. (2000) showed that flavonols, principally quercetin glycosides, accumulate in the surface cells of grape skins and other plant surfaces in response to sunlight or UV exposure. White and red grapes respond similarly. There they act as a UV screen by absorbing and converting UV wavelengths into harmless heat or light, while letting visible light through to the photosynthesizing and anthocyanin-bearing layers below. Quercetin amounts vary many-fold over the range of exposures, including within clusters, with responses right up to complete exposure to full sunlight. By contrast anthocyanins reach their maximum with only partial exposure, and in hot, sunny climates decline above that due to presumed degradation.

Consistent with these findings, Goldberg et al. (1998) showed several-fold differences among wine regions, with highest flavonol contents in warm and sunny climates (Australia, followed by California, South Africa and the south of France) and lowest in cool and cloudy climates such as of central and northern Europe, Canada and (surprisingly) Spain. They also differed, if somewhat inconsistently, among the grape varieties. Pinot Noir had very low contents in most regions, which helps to explain its weak wine colours.

The direct relationship of flavonols to berry sun exposure leads to important conclusions on cluster architecture. Loose clusters with small berries – whether for genetic or environmental reasons – must favour flavonol production. This gives a contributory explanation to why early post-setting moisture stress that limits berry size is positive for quality in red wines. There are also clear implications for clonal selection towards looser clusters in red grape varieties.

Information on the time course of flavonol accumulation is incomplete. Downey et al. (2003b), working with cvs Chardonnay and Shiraz in South Australia, noted two main periods during berry development as indicated by expression of the genes controlling the enzyme flavonol synthase: the first around flowering, and the second during late ripening after the main period of anthocyanin synthesis. On the other hand Price et al. (1995, 1996) found no particular relationship to phenological stage in Pinot Noir grown in Oregon, USA, and indeed some evidence of their degradation late in ripening under extreme temperatures. Measurements by Haselgrove et al. (2000) in South Australian Shiraz likewise showed quercetin-3-glucoside per berry as already high at veraison, with if anything a slight fall around maturity.

It thus seems most likely that flavonol synthesis occurs at any stage of berry development, depending on sun exposure, but perhaps especially pre-veraison. This would be consistent with the proven practice of lower-leaf defoliation starting at or soon after flowering to improve berry tolerance of heat and sunlight exposure later, at the same time potentially improving red wine colour and quality. Where quercetin has been measured, its content in the young wines has directly reflected that of the grapes.

A continuous potential for flavonol production through early berry development points to a relationship to maturity rankings paralleling that postulated for the tannins: that is, a pre-veraison increase proportional (other things being equal) to their differing durations of berry development from flowering to veraison. Less certainly, varietal differences in the duration of subsequent ripening might yield similar parallels, unless content reaches a saturation level or degradation sets in.

Together with some differences in the anthocyanins themselves, these relations help to explain the weak colours of red wines from the earliest-maturing grape varieties grown in the cool and cloudy climates to which they are suited, and the intense colours of those from most midseason and late varieties grown where they ripen fully and the fruit receives adequate exposure. Other types of varietal difference in the anthocyanins and flavonols presumably account for departures from the general rule.

Co-pigmentation explains the enhancement of colour (and perhaps palate qualities) of red wines co-fermented with small amounts of white grapes from certain varieties (Boulton 2001). The traditional Italian addition of Trebbiano and Malvasia to Sangiovese ferments is one such addition, as is that of Viognier to Syrah (Shiraz) in the French upper Rhône, and varieties such as Clairette, Roussanne and Grenache Blanc to Grenache etc. in Chateauneuf-du-Pape in the southern Rhône. And while the relative lateness of all the white varieties used may be necessary to make co-fermentation temporally possible, a further likely reason is that they provide the most co-factors. There appears to be little use of early white varieties in the same way, even though their early red partners should arguably have the most to gain.

10.5   Significance for the ecology of winemaking and drinking

Bringing all arguments together, some broad generalizations seem possible that link grape maturity groupings and their conditions of ripening to the best conditions for making and drinking the wines.

Early-maturing varieties, phenologically adapted to short and cool growing seasons, ripen best under cool conditions. Fermentation takes place best in the cool to cold autumns and winters of these climates. The low boiling point aromas that cool ripening and fermentation preserve are optimally expressed and perceived at low drinking temperatures, as are the wines' high natural acids and commonly retained balancing sugar. But tannins, as from late-maturing varieties, would be unpleasantly harsh on the palate at such wine temperatures.

By contrast the natural wines of late-maturing grapes, grown and fermented in warm to hot climates, are balanced for drinking at relatively high wine temperatures. These bring up the high boiling point aromatics that have survived warm ripening and fermentation, and soften perception of the high tannins. High acids at these temperatures would grate unpleasantly, while residual sugar (other than where balanced by high alcohol, tannins and flavour intensity as in fortified wines) tastes mawkish.

In between lies an infinite range of combinations and balances that express, and are best expressed in, their respective natural climates.

Here we see at work a long history of co-adaptation in grapegrowing, winemaking and wine drinking that has shaped not only the Old World's grape and wine industries, but also regional perceptions of wine style and quality. Northern Europe palates have naturally favoured the cool-climate styles of their region: a judgment long accepted as authoritative, because that was, until recently, the only major market for quality wines.

The growing production and consumption of quality wines in the New World, mostly in warm regions, together with the emergence (or renaissance) of quality viticulture in Mediterranean countries, is now bringing about a reappraisal of this opinion. There is new recognition that warm, and even quite hot, climates can give good wines with careful choice of grape varieties and the best terroirs for them. Mainly that means that vines are planted where temperatures are equable (lacking extremes), afternoon relative humidities high enough and the sun not too powerful. Mostly these terroirs are in coastal or near-coastal areas with moderating afternoon breezes. Some of the adapted varieties are already widely grown, but many more with potential must exist around the Mediterranean and Middle East.

Two factors have favoured the recent extensions of quality winemaking into warm areas. The first is improved winemaking technology, including night picking, must cooling and temperature-controlled fermentation, which improves the retention of aromatics and reduces other limitations to quality in these climates. The second is the spread of wine drinking more widely into warm-climate countries. Especially for red wines, drunk at room temperature, this favours wines produced in comparable climates.

The 21st century world market will see an extended range of quality wines, encompassing both cool and warm-climate styles. Adapted grapes will cover much of the maturity range. Achieving this will call mainly for exploration of quality potential among midseason and late-maturing grape varieties.

Chapter 11

Construction of Viticultural Climate Tables

This chapter describes the steps used here for constructing viticultural climate tables, which are slightly modified from those in *Viticulture and Environment* (Gladstones 1992). Chief differences are in the adjustment for diurnal temperature range in calculating biologically effective degree days (E°days), in the presentation of temperatures and sunshine hours, and in the inclusion of cloud cover and new temperature-related indices.

11.1   Preliminary appraisal

Recorded average mean temperature over the 'standard' vine growing season of April to October in the Northern Hemisphere, or October to April in the Southern Hemisphere, gives a preliminary measure of a site's or region's thermal regime for viticulture. Table 11.1 gives representative figures for some of the world's main wine-producing regions, together with their best-adapted grape maturity groups and typical wine styles. The matchings take into account the greater biological effectiveness of given mean temperatures at high latitudes (see subsection 2.1.3) and the fact that in very cool climates, vineyard sites are usually warmer than the standard climate data indicate.

Resulting seasonal averages are more accurate than using just the average mean temperature of the hottest month (Prescott 1969), or still more simply the average mean July or January temperature as is common in the recent literature. Both of these over-estimate available heat in continental climates and under-estimate it in maritime climates.

For sites or areas having no immediate climate data, growing season average mean temperatures can be estimated using sea-level temperature trends. Growing-season average means from surrounding weather stations are reduced to sea level by adding 0.6°C per 100 metres of their individual altitudes. Sea-level isotherms can then be drawn to give best overall fit. The

Table 11.1. Thermal regions for viticulture as measured by adjusted growing season average mean temperature. After Gladstones (2004).

Temp. °C	Wine styles	Typical regions
14–15	Very light and sparkling wines from earliest-maturing varieties	Southern England; Seattle, USA; Christchurch, New Zealand
15–16	Cool-climate still and sparkling wines from early-maturing varieties	Champagne, Burgundy, France; Rhine Valley, Germany; Willamette Valley, Oregon; Marlborough, New Zealand
16–17	Medium-bodied wines from early to midseason varieties	Loire Valley, France; Yarra Valley, Victoria; Coonawarra, South Australia
17–18	Medium to full-bodied wines from early to late-midseason varieties	Bordeaux, Northern Rhône, France; Napa Valley California; Barossa Valley, South Australia; Santiago, Chile
18–20	Full-bodied table wines from mainly midseason to late varieties; fortified sweet white wines from early varieties[1]	Southern France; North-eastern Victoria; Paso Robles, California
20–22	Full-bodied table wines from mainly midseason to late varieties[2]; fortified sweet red wines from mainly midseason varieties[1]	Douro Valley, Portugal; Riverland, South Australia; Swan Valley, Western Australia

[1] Fortified wines only with ample sunshine
[2] Quality table wines depending on sea breezes and/or cloud; bulk wines inland with full irrigation

process allows anomalous climate figures to be identified and discounted, e.g. high values due to urban warming, or unduly low means from recording sites in valleys or hollows where cold air ponding depresses the minima. The resulting sea-level isotherms allow estimation of land surface growing season averages for all points on the map, by first interpolating sea-level values, then subtracting 0.6°C per 100 metres of individual site altitudes.

Kirk and Hutchinson (1994) describe what is essentially the same exercise for a part of south-eastern Australia, using computer analysis to establish underlying latitudinal and longitudinal temperature trends and superimpose the effects of altitude. They expressed the result as a contour map of October–April simple °day totals over a 10°C base.

A problem with such methods is that while theoretically it is possible to apply them with any desired degree of spatial resolution, available information based on them is in most cases highly generalized. In Kirk and Hutchinson's study the temperature summation contours were based on calculated temperature averages for cells of $1/40 \times 1/40$ degrees, or approximately 2.2×2.75 kilometres. The DAYMET maps for California's Napa Valley cited by Swinchatt and Howell (2004) have similar limitations. Nor can users check the accuracy or appropriateness of the temperature data from which they were derived.

Beling et al. (2001) have shown that it is possible to incorporate factors for daylength and diurnal temperature range into the gridding process for mapping 'effective' isotherms and estimating vine phenology. However as their results show, the problem remains of discrepancies between modelled temperatures and those on the ground.

In any case none of the above procedures, no matter how accurate, can fully define viticultural climate. To be useful in practice this needs to include factors besides temperature such as rainfall, relative humidity, sunshine hours and cloudiness, and various measures of temperature extremes and variability. And as described in subsection 2.1.5, the effective temperatures experienced by vines can depend greatly on local factors such as slope, aspect and soil that gridded averages, and even directly altitude-adjusted temperatures, cannot entirely account for. That requires individual analysis for each site or terroir.

The longhand method that the rest of this chapter describes requires only standard, normally available climate statistics, adequately detailed contour maps, and a knowledge of local site characteristics. It employs only simple arithmetic and eye fitting of curves to monthly data.

The graphic method for estimating maturity dates and temperature averages for the final 30 days of ripening, to be described in Section 11.7, allows also the estimation of average ripening-period values for all other climatic elements for which monthly records are available, together with more detailed measures and indices of ripening-period temperature. Tabulating these gives a concise overview of climate as estimated for each vine maturity group during ripening.

11.2   Assembly and scrutiny of basic data

11.2.1 *Data recording periods*

Ideally climatic comparisons should be based on long-term data, to iron out short-term (seasonal to decadal) fluctuations that occur in all climates. Thirty years of records is the commonly accepted standard for reliable comparisons. Many officially published figures are as 30-year 'normals', of which the most widely published for the 20th century have been those for 1931–60 and 1961–

90. As far as possible I use temperatures representing the second half of the 20th century for table construction and inter-site or inter-regional comparisons. The 1961–90 normals fairly represent the period. Nor should inclusion of the 1931–60 normals or other data back to 1930 greatly change the picture, since average temperatures for the two periods hardly differ. However inclusion of the unusually warm 1990s will appreciably bias recent short-term records, while most pre-1930 figures, including the 1901–30 normals, must be excluded as being from an unusually cold period.

How far the climate data from the mid to late 20th century can be expected to represent future conditions will be examined in Chapters 12 and 13. For now we may merely note that the evidence, properly analysed, does not suggest any large climate change or viticultural shifts over the next few decades. Also to be noted is that the tables as presently calculated calibrate temperature records from the second half of the 20th century against vine phenology as observed in that period. The calibration, as such, should not change with temperature movements in either direction

11.2.2 *Local influences on temperature records*
The topographic locations of recording stations need to be allowed for wherever possible. Many long-established stations are in valleys and may have obstructed air drainage. Ponding of cold air at night then depresses the minima and means below the countryside average (unless offset over time by urban warming: see below). In contrast slopes with good night air drainage, such as might be selected for viticulture, usually have minima and means above the countryside average for the same altitudes. The effect is strongest on isolated hills and projecting ridges, as discussed in Section 4.3.

In the absence of topographic information these things can be roughly inferred from the climate data themselves. A wide diurnal range, for the region, indicates poor night air drainage and below-average means, and a narrow range good night air drainage and above-average means.

The other main distorting factor is urban warming, when recording stations are in or near towns or cities. This artificially raises the minima and may slightly lower the maxima, due to daytime absorption of heat by brick or stone buildings, roads, pavements etc. and its re-radiation through the night. Energy input from artificial heating and cooling of buildings, industry and motor transport further heats urban localities. As well air pollution may reduce sunshine access during the day but traps outgoing radiant heat at night. The combined result is always higher minima and means and a reduced diurnal range.

Table 11.2. Effects of urbanization according to population on the USA
temperature record 1901–1984, relative to centres with populations <2,000.
After Karl et al. (1988).

Population	Minimum °C	Maximum °C	Average °C[1]	Diurnal Range °C
2,000	+0.12	−0.01	+0.06	−0.13
5,000	+0.16	−0.02	+0.08	−0.18
10,000	+0.22	−0.02	+0.11	−0.24
20,000	+0.32	−0.03	+0.16	−0.35
50,000	+0.48	−0.05	+0.24	−0.53
100,000	+0.63	−0.07	+0.32	−0.70
200,000	+0.87	−0.09	+0.44	−0.96
500,000	+1.33	−0.14	+0.67	−1.47
1,000,000	+1.81	−0.20	+0.91	−2.00
2,000,000	+2.48	−0.27	+1.25	−2.74
5,000,000	+3.73	−0.40	+1.88	−4.12
10,000,000	+5.10	−0.55	+2.57	−5.63

[1]　Karl et al. show these figures as averages, not means, but that they were calculated as
(min. + max.) /2. The figures themselves suggest that the authors wrongly disregarded
urban effects on the maxima. Correct means can be estimated from the minima and
maxima.

The intensity of urban warming increases with town or city size as
roughly indicated by population, though with some influence of layout and
architecture. Karl et al. (1988) reported a detailed study of urban warming
in a network of 541 stations across the USA for the period 1901–84 and a
further 941 for 1941–84, using records from towns with 1980 populations of
less than 2000 as the baseline. Table 11.2 summarizes the results. This work
made clear that even small towns have some local effect on the temperature
records. In the case of large centres there must also be some carry-over to the
immediate surrounding countryside through the air convections their heat
islands set up.

Karl et al. emphasized that their results do not necessarily apply outside
the USA, owing the differences in town/city structure and energy use.
Sturman and Tapper (1996) presented evidence that European cities show
similar but slightly lesser warming relative to population size, and Australian

and New Zealand cities a little less again: differences probably related to their relative densities of population and buildings and intensities of energy input. The phenomenon is nevertheless clearly universal, as indeed has been common knowledge in Europe for centuries. Comparisons I have been able to make as part of this study for sites in and around London (England), Paris (France) and Auckland and Christchurch (New Zealand) have been fully consistent with the other findings cited, with the heat islands at least of major cities extending well beyond their city boundaries and comparable lesser effects for smaller centres.

Selection of records for viticultural prediction must therefore favour those from small to, at most, moderate-sized centres. Many of these will have minor urban warming, but that is broadly subsumed in the calibration against grape maturity groups (Table 10.1), which unavoidably depends on what relatively rural climate records have been available.

11.3   Temperature adjustments for location and altitude

Where vineyard sites are close to a reliable weather station with long-term records, or align closely with it in their mapped sea-level isotherms, no adjustment for location is needed other than for differences in altitude (0.6°C per 100 metres) and any other site differences as described in subsection 2.1.5 and Section 11.5 below. Ideally these adjustments are made for the individual growing season months, but for most purposes growing season averages suffice.

Where a site is at some distance from its nearest recording station, and not on the same mapped sea-level isotherm, the sea-level difference can be estimated by interpolation. The difference, combined with that for altitude, is then applied to the monthly station records, assuming for practical purposes that it applies equally to the minima, maxima and means. Alternatively serviceable estimates may be possible using composites from multiple surrounding stations.

Since the 1940s and 1950s, city weather stations have increasingly relocated to airports and airfields, which by their nature are mostly flat and open; while most other stations now accepted as giving valid and publishable results occupy as neutral terroirs as are conveniently available. Some exceptions remain, such as lighthouse sites that are anomalous through being both strictly coastal and usually located on hilltops, and those still located within cities or large towns. But barring these, adjustments for vineyard sites can now usually be made directly from uncorrected station data without major error.

Site estimates for other climatic elements besides temperature can be interpolated from surrounding records with mostly satisfactory accuracy; differences in them are in general less critical than those of temperature, because they do not affect vine phenology. Rainfall and relative humidity relate oppositely to temperature with altitude, which must therefore be taken into account, as must closeness to coasts. On the other hand sunshine hours and cloud vary little with altitude over the restricted range of most viticultural comparisons.

Certain measures of the variability of minimum and maximum temperatures can be useful in estimating monthly average lowest minima and highest maxima where these are not directly available, as is often the case. Subsection 11.5.1 will describe their practical use for this purpose.

11.4   Temperature adjustments for slope, aspect, soil, water and wind

Subsection 2.1.5 and Table 2.2 summarize the suggested temperature adjustments for these factors in estimating E°days. Section 3.4 gives a more extended treatment of wind effects, and Sections 4.3 to 4.7 those of slope, aspect, soil and proximity to oceans, lakes and rivers. Here we need only to reiterate two general points.

The first is that all site and diurnal temperatures range adjustments are additive, with minima and maxima treated separately.

Secondly, the suggested adjustments in Table 2.2 can obviously be no more than guidelines, which have to be interpreted for individual terroirs. This is an unavoidably subjective process, but carefully done it undoubtedly improves predictive power. The site analysis that it entails is in any case a necessary part of any serious preparation for vineyard planting.

11.5   Tabulating the basic data

The upper halves of the tables record the basic data from source recording stations, without adjustments except in the three columns showing the progressively estimated E°days.

It is essential that recording station latitude, longitude and altitude be precisely specified. This allows correct adjustments for the location and altitude of vineyard site(s) being estimated for, and for any alternative sites in the future. Similarly the period of temperature records used needs to be specified to establish comparability with other records. (Not all published records are fully reliable in this respect, with some '30-year normals' being from shorter periods judged to be representative. Nor are changes of thermometer site or method of temperature recording always identified. However most modern published records are reasonably accurate.)

11.5.1 *Temperatures and temperature indices*

The column of lowest recorded minima can be a useful guide for viticulture but has to be interpreted with caution, for two reasons. Firstly, the data come only from the specified recording period, which may be too short or not representative enough to indicate ultimate risk. Secondly, they too easily stem from uncorrected mistakes in the record: it takes only one! Monthly average lowest minima (available or estimated as described below), and especially the spring frost indices that can be derived from them, are better buffered against error and give a more reliable guide to frost risk.

Records of average monthly lowest minima and highest maxima are seldom included among published climate data, which is a pity because they are biologically important. However with modern data recording and processing they are readily enough extracted where needed. For instance the Australian Bureau of Meteorology does not include them in standard climate sheets but can provide them on request. The US National Climate Data Center (NCDC 1996, CD-ROM) likewise does not tabulate such averages, but they can be calculated from the data provided in many cases.

Where records of average monthly lowest minima and highest maxima are available for some stations in a general region, it is usually possible to estimate them for the others with fair accuracy. That is because the monthly differences between average minimum and average lowest minimum, and between average maximum and average highest maximum, are relatively constant across regions of broad climatic similarity. I refer to these as V_{min} and V_{max}, being measures of the variability of the minimum and maximum respectively. Local topography, while partly governing absolute values of the minima and maxima, has little influence on their variability, which depends more on the behaviour of regional wind systems.

Once V_{min} and V_{max} have been established for some site(s) in a natural geographic region they can be applied as needed to related sites having data only for the monthly average minima and maxima. Some degree of error is acceptable here, because the resulting figures are general indicators only and do not contribute to estimating maturity dates.

The *Spring Frost Index (SFI)* for each month is calculated simply as the average mean minus the average lowest minimum (for further discussion see subsection 2.4.2). I calculate these data only for the spring to early summer months in which some possibility of frost remains; or where that is negligible, in any case for the first two months of the growing season. The same calculation for winter months is also useful in continental climates to estimate the contribution of temperature variability to winter killing.

The *Heat Stressfulness Index (HSI)* is a month's average highest maximum minus its average mean. It has significance throughout the growing season, but especially at veraison and during ripening when irregular heat spikes are most damaging.

The *Temperature Variability Index (TVI)* is most easily calculated as the sum of the average maximum and average highest maximum, minus the sum of the average minimum and the average lowest minimum. It provides a general indication of temperature variability for the growing season or individual parts within it; while the average for the non-growing season months, together with that for the SFI (above), illustrates the contribution of short-term temperature variability to winter killing.

11.5.2 *Sunshine, cloud, rainfall and relative humidity*

The available data on *sunshine hours* pose problems of interpretation and reliability. Mostly they are cited as 'hours of bright sunshine', but the definitions of bright sunshine are unclear and the historical methods of measurement mostly crude. Sparsity of recording sites seems to be universal, with the records from not a few contaminated by local factors such as shadowing by hills. The sunshine maps derived from these data must therefore be treated with caution, as must spuriously precise-looking monthly figures interpolated from them or derived from computer models.

That said, the need for fine precision is not great in the present context. Available estimates are probably adequate for the terroir comparisons we seek. Also it is safe enough to interpolate for new sites within regions, provided the contrasts in altitude and/or rainfall are not excessive. Sunshine hours are relatively unresponsive to these factors over the ranges normally encountered in viticulture.

The raw data for *average cloud cover*, expressed here in oktas, or eighths of the sky covered, come from far more comprehensive (if subjective) recordings. They provide a useful complement to sunshine hours, since although the two are inversely related, they are not completely so. Ratings of cloud cover have long been part of the standard observations at climate recording stations, although the reported scales and timing of recordings differ among countries. Australian daytime recordings are in oktas at 9 a.m. and 3 p.m. Figures from outside Australia are reported mostly as averages of two or three daylight times, or as unspecified daytime averages. Probably the timing differences do not matter much, since in typical viticultural climates average cloudiness appears to stay fairly uniform through the day. (An exception is the sub-tropics, where summer days tend to start clear, but cloud builds up in the afternoon.) The figures I use are daytime averages as best they can be derived. As with sunshine hours these are probably adequate to depict cloudiness as a terroir component. And as with sunshine hours, estimates by interpolation can be made with some confidence within regions.

Virtually all sites with temperature records also have *rainfall* records, often for longer periods. Desirably the periods used for temperature and rainfall should be the same; or if the available temperature records are shorter than optimal, a longer rainfall record is better if it still typifies the period in focus (in this case the mid to late 20th century).

Estimating rainfall by interpolation calls for care, due to the marked local influences of altitude and exposure to rain-bearing winds. In most regions reasonably detailed rainfall maps are available that can give guidance.

The published records of *relative humidity* raise difficulties that can be only partly resolved, but need to be understood because of their importance to terroir (see Sections 3.3, 6.1). The most immediate is that the timing of those representing the afternoon differs widely among countries. Some, including Australia, specify 3 p.m., which is probably the most useful for present purposes because it is close to the hottest time of day. Together with the maximum temperature it gives a good indication of potential moisture stresses. In near-coastal areas it also gives some measure of the prevalence and strength of afternoon sea breezes. Other published timings range between noon and 4 p.m. and are to varying degrees not comparable. Thus whatever the timing of the data available and used, it needs to be clearly specified.

11.6   Estimating effective degree days (E°days)

11.6.1 *Stages for tabulation*
The tables as presented here (Appendix 2) show estimation of E°days in three stages, the first with just the 19°C mean temperature cap over a 10°C base. It and the subsequent steps include, where necessary, special treatment for low October/April temperatures as detailed in subsection 2.1.4 and Figure 2.3. The second stage has additional adjustments to the raw data for diurnal temperature range and daylength, which apply universally. The third stage then takes in local factors of location, topography etc. for projected individual or typical vineyard sites. Staging in this way highlights the relative contributions of the different adjustment factors. For the theory underlying them see Section 2.1 and, more broadly, Chapter 4.

11.6.2 *Working through the adjustments: example of Canberra, Australia*

Table 11.3 illustrates the workings to estimate and tabulate E°days for Canberra, ACT (Australian Capital Territory). Base data are for the airport, which is on flat land a little east of the city. Because of its clear separation from the city and the fact that Canberra is a relatively small centre (1990 population c. 300,000) with no tall buildings, urban warming effects can be considered negligible. Airport latitude is 35°18'S, altitude 578 metres, and the temperature record is for 1957–2000.

Part 1 of Table 11.3 lists the site factors to be ultimately allowed for, taking the vineyards around the nearby village of Murrumbateman, New South Wales, as representative of the region. Sea-level temperature trends for the growing season (Gladstones 1992, Figure 14) suggest the Murrumbateman vineyards, about 30–40 kilometres north of Canberra Airport, to be some 0.2°C warmer at comparable altitude. The vineyards are mostly 50–100 metres higher than the airport, on moderate slopes of no particular aspect, some on isolated hills or ridges. Because the airport weather station is on flat land there is full adjustment for typical vineyard slope. All Murrumbateman vineyards are inland from the main divide of the Great Dividing Range, largely isolated from summer marine winds from the south-east and exposed more to warm winds from the inland. Soils are predominantly shallowish brown clay loams (Halliday 2006), not obviously requiring adjustment for soil type. No lakes etc. are nearby.

Part 2 of Table 11.3 carries forward the raw station temperatures to include adjustments for diurnal range and latitude as well as 19°C capping, but not yet local site factors.

(a) Column 2 lists the raw mean temperatures for each of the growing season months.
(b) Column 3 lists the raw diurnal ranges, but in this case for October, November and December only. January temperatures, to which the range adjustment might otherwise apply, are clearly above the 19°C cap and so do not need calculating; while February, March and April are beyond the time when diurnal range counts: see subsection 2.1.2.
(c) Column 4 shows the adjustments to the means for the applicable monthly diurnal ranges: in this case the °C excesses of range over 12.0, multiplied by –0.25.
(d) Column 5 gives the latitude (daylength) multiplier for each month as estimated from graphs based on Table 2.1. Again these are not needed for January and February because these months' finally adjusted effective

Table 11.3. Calculation of biologically effective degree days (E°days) for Canberra district, Australia.

1. Adjustments to raw Canberra data for typical vineyard sites, °C

Minimum	Maximum	
+0.2	+0.2	Vineyards 30–40 km N
−0.4	−0.4	Average altitude 650 m
+0.6	−	Moderate slopes, some isolated hills
−	+0.4	Moderate isolation from cool winds
+0.4	+0.2	Mean +0.3°, range −0.2°

2. Adjustments to raw Canberra data for diurnal range, latitude and 19°C cap

	Average mean °C	Average range °C	Range adjustment °C	Latitude factor	E°days
October	12.8	13.1	−0.275	0.987	77
November	15.7	13.9	−0.475	0.976	153
December	18.6	14.4	−0.600	0.970	241
January	20.4	−	−	−	279
February	20.1	−	−	−	252
March	17.5	−	−	0.996	232
April	13.3	−	−	1.013	100
October-April					1334

3. As 2 plus vineyard site adjustments

	Average mean °C	Average range °C	Range adjustment °C	Latitude factor	E°days
October	13.1	12.9	−0.225	0.987	88
November	16.0	13.7	−0.425	0.976	163
December	18.9	14.2	−0.550	0.970	251
January	20.7	−	−	−	279
February	20.4	−	−	−	252
March	17.8	−	−	0.996	241
April	13.6	−	−	1.013	109
October–April					1383

temperatures clearly exceed 19°C; but they apply to all other months.

(e) Calculation of E°days for October is then as follows: 2.8 (i.e. 12.8 minus 10), minus 0.275° diurnal range adjustment, multiplied by 0.987 latitude factor, multiplied by 31 days = 77 E°days.

Other months follow in the same way. Where final 'effective heat' for a month equals or exceeds 9°C (effective mean temperature 19°C), E°days equal the number of days in the month × 9.

For the final site-adjusted calculation (Part 3 of Table 1.3) the process of Part 2 is repeated, using changed average means and diurnal ranges according to the net sums of the site adjustments shown in Part 1. The latitude factors remain the same.

If estimates are required for Canberra-related sites other than Murrumbateman, only Part 3 of the calculations needs to be changed using appropriate new location and site adjustments to the minima and maxima. The same applies to projected climate changes. Site and climate changes can be combined in the same calculation if desired.

It is coincidental that in the case of Canberra/Murrumbateman the final E°days differ only a little from the raw °days with 19°C cut-off, which are simply calculated to be 1388. That is not uncommon for inland low-latitude vineyards, where selection of the locally more desirable terroirs can more or less exactly offset the phenologically negative effects of low latitude and regionally wide diurnal temperature range.

It is different at high latitudes. Usually, even inland, these have narrow diurnal ranges and attract positive temperature adjustments for both diurnal range and latitude. Additionally, vine survival and fruit ripening often demand most rigorous selection of the warmest and most equable slopes and soils. E°days there for vineyard sites are commonly 100–200 above the raw figures from typical non-city weather stations, with some up to 250 or even 300 E°days warmer when adjustments for latitude, diurnal range and site selection are combined. Such differences are obviously crucial for viticultural climate estimation at the cool margin.

Note that whereas the Canberra example shows full workings for the tables as constructed here, in practical use all adjustments for diurnal range, latitude, site and final 19°C capping can be combined in a single calculation by including site adjustments from the beginning.

11.7 Predicting maturity dates and ripening conditions

Here we carry forward the process described in Section 11.6, using the monthly accumulations of E°days to fill in the bottom halves of the tables. These show firstly the predicted average ripening dates for each (if they ripen) of the nine grape maturity groups, and from them their estimated average conditions of temperature, sunshine, rainfall etc. for the preceding 30 days.

The following description continues to take Canberra, Australia as its example. I use graphic methods throughout and find it convenient to use 400 × 600 mm graph sheet gridded at 5 mm intervals. For most climates this gives a big enough horizontal span to cover the full range of ripening dates at one day per division and a vertical scale that accommodates the range of ripening-period temperatures at 0.1 or 0.2°C per division. The sheets can be extended if necessary, or the scales halved.

11.7.1   *Average maturity dates for maturity groups*

Step one is to graph the estimated monthly average (i.e. mid-month) site mean temperatures, in this case the recording station means plus 0.3°C, over the potential range of ripening dates and extending back at least 15 days before that estimated for Maturity Group 1. For Canberra that was easy to establish because the accumulation at the end of February was 1033 E°days and the requirement 1020 E°days (Table 10.1). With nine E°days per day because average means at this time still exceed 19°C, this firmly places the maturity date at February 27 and the beginning of the graph no later than February 12. A temperature curve is then drawn from that date giving best fit to the graphed mid-month means to the end of April (October in the Northern Hemisphere), or until E°days have reached 1500, whichever comes first.

Estimating maturity for the following maturity group in this example makes a new start with the beginning of March, i.e. from 1033 E°days. Effective means start to fall below 19°C during this month, so there is a progressive fall in E°days per day and the estimate of accumulation must progress incrementally by projecting forward along the temperature curve. I do this in blocks of three days. Here the requirement of 1080 E°days for Maturity Group 2 is estimated to be reached after 5.22 days (March 6), with the succeeding 60 E°day intervals needing 6.98, 8.00 and 9.09 days respectively as shown in Table 11.4. The last date so estimated, giving March 30, also happened to agree with that estimated by working back from the end of the month. For April in this case there was no need for any such process because the requirement for Maturity Group 7 was met clearly on the last day of the month.

Table 11.4. Estimating average grape maturity dates, Canberra district, Australia.

E°days at end of	Maturity group and E°day requirements		Estimated date reached	Days for next 60 E°days
January 781	1	1020	February 27	
February 1033	2	1080	March 5.22 (March 6)	6.98
March 1274	3	1140	March 12.20 (March 13)	8.00
April 1983	4	1200	March 20.20 (March 21)	9.09
	5	1260	March 29.29 (March 30)	
	6	1320	April 9.39 (April 10)	
	7	1380	April 30	
	8	1440	–	
	9	1500	–	

A source of error in the calculations for some sites needs to be recognized. Where the curve of effective temperature passes through 19°C during any calendar month, the E°days estimated for that month can differ depending on whether it is based on the temperature curve, or on the month's average mean temperature as in the upper part of the table. Thus if the average effective mean is exactly 19°C, the month will receive maximum E°days. But working from the curve, some E°days will be lost from the part of the curve above 19°C that will not be made up in the part below it, resulting in less total E°days for the month. Mostly the discrepancy is negligible. The exception is in warm to hot continental climates, where temperatures during ripening both cross 19°C and change steeply with time. I prefer, for the sake of simplicity, to accept the resulting error and, where necessary, adjust the estimates of ripening dates through that month to conform better to the E°days derived from the monthly average means. This allows a smooth transition to a fresh start of maturity estimation from the beginning of the following calendar month, which will not be affected in the same way.

Note that, as discussed in subsection 2.1.2, diurnal temperature ranges do not contribute to estimating E°days at this time of season. Latitude/daylength do, mainly at high latitudes; but since ripening there is around the equinox, when all latitudes have the same daylength, any contribution through ripening is negligible. Both factors can therefore be safely disregarded in these calculations. Only the individual site adjustments to the mean are needed.

There is little point to pursuing ripening dates beyond the end of April/October, because in nearly all regions a later average ripening date is too precarious for economic viticulture. However dates may, if desired, be projected into early the following month for the rare cool climates with reliably dry autumns.

11.7.2 *Average conditions for the final 30 days of ripening*

Having established the curve of effective temperatures over the full range of potential ripening periods, and estimated maturity dates for those maturity groups that will ripen, it is a simple matter to estimate averages of temperature and of all other climatic elements for the final 30 days of ripening for each maturity group.

The first step is to measure back 15 days from the respective estimated ripening dates, and there rule a vertical ink bar for each maturity group. That is the reason why the graphs must start at least 15 days before the maturity date for Group 1. The bars should extend to the full height of the graph paper. Their intersections with the temperature curve give an adequate estimate of average effective mean temperature for the final 30 ripening days of each maturity group that ripens.

Now the graphing process can be repeated on the same sheet for all other available climatic measures, in erasable pencil if desired and using whatever vertical scales fit best. In the Canberra example the average lowest monthly minima and average minima will both be raised by 0.4°C compared with the raw station data to accommodate the net site adjustment, and the average maxima and average highest maxima by 0.2°C. Similarly the raw HSIs will be reduced by 0.1 for graphing, and the raw TVIs by 0.4. Sun hours and cloud seldom need site adjustments, but they may sometimes be called for in respect to rainfall and especially relative humidity, for instance where recording and vineyard sites differ markedly in their closeness to the coast.

How accurately do E°days predict average maturity dates? Obviously precision can never be complete, given the approximations involved and that seasons and even climate vary to some extent. But generally the fit is surprisingly good. The main weaknesses are at the climatic extremes.

In very warm, maritime climates (examples I am familiar with are the Swan Valley in Western Australia and the Hunter Valley in New South Wales), early-maturing varieties such as Chardonnay ripen perhaps 10 days earlier than predicted. That is because: (1) counting the season only from 1 October misses prior warm weather; and (2) these varieties mostly have weak winter dormancy, which contributes further to their early budbreak and following phenology. But the consequences for the ripening period are small. Because early varieties here ripen in mid summer, when average temperature changes little from week to week, differences in ripening time have little impact on their ripening conditions. Midseason and late-maturing grape varieties in the same climates appear to ripen much as predicted, as do most varieties in hot but less maritime climates. Thus there is little case for changing a defined start of season that otherwise works well and has become standard in the literature.

At the cool limit of ripening, particularly in maritime climates, the slow final accretion of E°days makes end-of-season date estimates excessively extended and sensitive to small temperature differences. Autumn rainfall and frost allowing, ripening may proceed intermittently through occasional warm and sunny spells over a long time. Thus the latest estimated ripening dates in these climates are at best very approximate. Cool climates with a dry and sunny late autumn, such as in Central Otago, New Zealand, or the Derwent Valley in southern Tasmania may fairly regularly be able to ripen grapes a maturity group or so later than the temperature averages would suggest. The variability among seasons, and in some cases the dominating role of autumn rains in defining growing season length, nevertheless means that in many environments it is prudent to allow a safety margin of at least one to two maturity groups to ensure satisfactory ripening.

11.8   Some final points on the tables

The usefulness of the final tables will depend on situation. However a few general points can be made.

The first is that rough allowances are possible for wine styles other than dry, still table wines. Grapes for sparkling wines are normally picked earlier, needing a ripening capacity (or picking date) less by the equivalent of one to two maturity groups. Similarly red grapes for still rosé wines will normally be picked at least a maturity group earlier than for red wines (tannin maturity is not a concern here because the ferment is off skins before significant tannin extraction occurs). Sweet fortified wines, on the other hand, need a site ripening capacity at least one to two maturity groups more than for still table wines. There is no general rule for sweet table wines, but ideally they need mild ripening to normal maturity, followed by cool, dry but humid weather to allow sugar and flavour concentration under botrytis infection while conserving natural acids and volatile aromas.

A second point is that it remains open how far grape varieties of comparable maturities can differ in their optimum ripening temperatures. The common adaptation of Riesling and Sauvignon Blanc to cool ripening climates is a case in point. Yet Riesling in Australia also gives some of its best wines, if of different style, in moderately warm climates whereas Sauvignon Blanc does not. Verdelho, a variety of similarly early ripening to Sauvignon Blanc, can give excellent wines in hot climates such as that of Western Australia's Swan Valley. Presumably these differences are because varieties differ in their distinctive suites of aromatics, in the individual responses of these to ripening temperature, in varietal capacities to store them in the berries as stable and non-volatile conjugates (see Section 2.3), and in berry resistance to heat injury.

That red varieties grown for red wines need more heat than white varieties for full ripening is well established (Table 10.1). At the same time the greater berry absorption of heat because of their colour makes them potentially more susceptible to both heat injury and sub-clinical quality loss. This is most evident in early-maturing red varieties such as Pinot Noir: the result being that they have a notably restricted range of optimal climates and need great temperature equability. By contrast the late-maturing varieties grown in warm to hot climates, giving full-bodied and mostly tannic wines, generally tolerate temperature variability and extremes. This could be, as Section 10.5 argues, because the prolonged Stage 2 of berry development allows a substantial build-up of tannins and flavonols in the skins.

The question then is: do the mostly midseason to late-maturing varieties grown in hot climates, both red and white, need their accustomed warm temperatures for best ripening? If so their ripening conditions as estimated in the climate tables need to be interpreted accordingly.

Historical experience of climate and wine styles, as summarized in Table 3.1 and developed further in Chapter 10, tends to support this idea. So does an observation that leads to a convenient rule of thumb. That is, that across a very wide range of climates the best adapted varieties (for table wines) are those that reach maturity in the second half of the first month of autumn, i.e. of September in the Northern Hemisphere and March in the Southern Hemisphere.

Temperatures at these times and the resulting wine styles differ markedly across environments, but optimum ripening dates stay broadly the same. In the climate tables the half-month span forms a window, through which can be read the likely best-adapted maturity groups. The presence of such a window that ripens quality grape varieties is one measure of an environment's suitability for viticulture.

Chapter 12

Climate Change

The intimate connection between viticulture and climate has three important aspects relating to climate change. First, any climate change must obviously contribute to shaping future viticultural terroirs. Given that the vines have a commercial life of 50 or even 100 years, any reliable pointers to future climate must play a crucial part in vineyard planning.

Second, historical accounts of vine growing, and of its geographical limits and vintage dates, give valuable insights into past climates. Elucidation of these helps us to estimate how much of present and future climate change might be due to natural causes, since what has happened naturally before can, and almost certainly will, happen naturally again. Only against that background can we start to assess any new contribution of man-caused, or anthropogenic, factors.

Finally, our already-described studies of vine phenology in relation to topography and other land surface characteristics (e.g. Table 2.2) give a useful but hitherto neglected approach to estimating errors in the thermometer record of land surface temperatures. Dating from about 1850, this contains much potential for error and bias that has not been properly assessed.

This chapter falls into two parts. The first (Sections 12.1–12.2) deals with pre-industrial and pre-instrumental climates as inferred from a variety of surrogate measures, including those from the records of European viticulture and vine phenology.

The second (Sections 12.3 on) examines mainly post-1850 measured temperatures, and inferences from them on both natural and anthropogenic climate change. Such a study is central to the climate debate. If the temperature record is wrong or biased, so will be the models and attributions built on it. Indeed, small initial biases can easily become magnified. That is a part of climate change studies where I think core researchers have signally failed, but where the insights we have gained from studying viticultural climates, and especially mesoclimates, can make a new and useful contribution.

12.1   Pre-industrial climates

The surrogate measures that must serve to indicate pre-instrumental temperatures take a variety of forms. They include ocean levels as measured by tidal gauges, the height and dating of coastal sedimentary benches and coral, and the history of shallow coastlines and estuaries; the seasonal extents of sea ice, mainly around Iceland and Greenland, and effects on shipping and colonization there; the advance and retreat of glaciers; agricultural records, not only of grapevines but also of other horticultural and field crops with known climatic requirements; records of major volcanic eruptions, producing dust veils that reflect away sunlight and cool the earth for up to several years; and records of observed sunspots and mid-latitude auroras, which we now know to be correlated with solar output and the solar magnetic field.

Other, more recently studied correlates of the sun's energy output and magnetic field include inversely varying atmospheric levels of the carbon isotope ^{14}C, which becomes laid down and measurable in tree rings that can be exactly dated; the thickness and densities of the tree rings themselves where tree growth is limited by low growing-season temperatures (as at high latitude and mountain upper treelines); an inverse relationship of solar activity with the atmospherically produced beryllium isotope ^{10}Be, laid down in polar ice sheets with dateable annual snowfalls; and direct measures of fluctuating historical temperatures as indicated by temperature anomalies in deep bore holes.

12.1.1   *The evidence of sea levels and early history*

The work of Fairbridge (1962) on world sea levels was a notable early contribution to climate change studies which has been inexplicably overlooked in the more recent literature. Based on tidal gauges going back some 300 years and coastal geomorphology before that, he concluded that within the last 1200 years the sea twice exceeded its mid 20th century levels: by about 20 cm in the 9th and early 10th centuries and by up to 40 cm between the mid 11th and early 13th centuries (see Figure 12.1). The latter peak coincided with what Fairbridge referred to as the 'Post Carolingian', or 'Rottnest' submergence, and gave evidence of global temperatures at the time significantly above those of the 20th century. Geomorphology then indicated a progressive sea level fall of a metre or more to a minimum in the 15th century, followed by a steep recovery to modern levels though the early to mid 16th

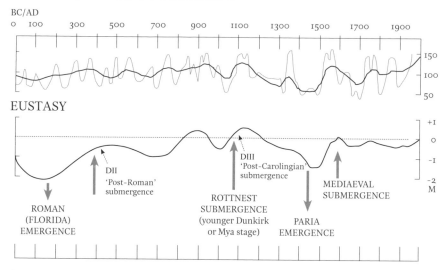

SUNSPOTS (Schove) 30-year smoothed maxima

Figure 12.1. World sea levels (eustasy) and sunspot records of the Christian era. From Fairbridge (1962), including sunspot data of Schove (1955). Sea levels as metres above or below those of the mid-late 20th century.

century. Fluctuations over the subsequent tidal gauge period have been more subdued: a fall of perhaps 20–30 cm though the 17th century, followed by a rise to near-present levels that persisted though much of the 18th century, renewed fall of up to 20 cm in the early 19th century, and a mostly steady rise from the mid 19th century on. The rate of rise through this last period has averaged about 1.2 mm a year and has shown reasonable correlation with measured air temperatures (Gornitz et al. 1982). As Figure 12.1 also shows, there is good general agreement between Fairbridge's sea levels and the historical records of sunspots as documented by Schove (1955).

Fairbridge made the important point that, unlike many measures, sea level gives a true indication of world average temperatures because of the hydrostatic continuity of the world water plane. He also underlined the correlations of climate and sea level with solar phenomena, as the following passage from his 1962 paper shows.

The solar constant … is very stable, but its ultraviolet emission varies over 200% with the sunspot cycles; this in turn controls the density of the upper atmosphere ozone screen which acts as a thermal blanket to the earth. Furthermore the corpuscular cosmic 'ray' streams associated with sunspot periods also play a role.

Thus past records of sunspots should reflect climatic patterns. Observations of sunspots and aurora back to 649 BC (Schove 1955) give a pattern that is remarkably reflected by the climatic records, historical and geological. Sea levels, in turn, react to the climatic events with remarkable clarity. The significant feature about the effect of sunspot cycles on climate is that the short 11-year cycles play a role that is too short to show a very great sea-level change, only a few cm as a rule. But, if one smooths the sunspot peaks over 100-year periods, the oscillation is almost exactly matched by the general record of mean sea-level.

The most comprehensive other historical and geographical studies of climate over the last two millennia, albeit mostly for Europe, have been those of Le Roy Ladurie (1971), Lamb (1977, 1982, 1984) and Pfister (1984, 1988). Bradley (1991) also gives a useful summary.

These studies present a picture broadly consistent with Fairbridge's sea levels, particularly for the last millennium for which the information is strongest. The most prominent feature is the marked warmth of the so-called Mediaeval Warm Period, lasting (with some fluctuation) from the ninth to about the late 13th century. During this period sea ice was largely absent from around Iceland and southern Greenland. Viking settlers grew cereals in both regions, and in Greenland they buried their dead deep in soils that later became permanently frozen. Cereal cultivation in Scandinavia extended much further north than now, with wheat grown in the Trondheim district and hardier cereals beyond 69° latitude. Throughout western Europe crop cultivation and treelines went some 200 m higher than in modern times. Glaciers were substantially less extensive than they were to become later, while coastlines corroborated Fairbridge's geomorphologically-derived estimates of sea levels at the time.

12.1.2 *The evidence of historical viticulture*

Records of mediaeval viticulture paralleled those of other crops. European wine production reached the Baltic coast as far east as East Prussia. According to Weinhold (1978), the more southerly parts of Holstein, Mecklenburg, Pomerania and West and East Prussia continued to be flourishing wine regions into the 15th and 16th centuries.

Debuigne (1976) noted that viticulture, the first since Roman times, became re-established in Belgium in the ninth century and was still present in the following two centuries. Hyams (1987) reported that the period 1150–1300 saw increased planting there, as well as in Luxembourg and England.

Unwin (1990) made a close study of the records relating to Saxon and early Norman viticulture in England, and concluded that the evidence for pre-Conquest viticulture on any scale was limited and equivocal. At most it was probably confined to monastic institutions making wine for their own use. But the coming of the Normans changed all that. By the time of the Domesday Book (1086), viticulture was widespread over the southern third of the country. Most appeared to be in secular hands, with vineyards especially situated on prosperous manors and up to several acres in size. The more comprehensive available evidence from the 13th century showed vines continuing to play an important role in the manorial economy.

The map of Lamb (1977) of English vineyards between AD 1000 and 1300 likewise shows vines over the southern one-third of the country, with a northern latitude limit of about 53°. On the qualities of the wines little is known, but they must have been competitive enough with imports from France. Lamb records one quote from William of Malmesbury, written about 1150. Describing the Vale of Gloucester, he wrote: "No county in England has so many or so good vineyards as this, either for fertility or for sweetness of the grape. The wine has no unpleasant tartness or eagerness; and it is little inferior to the French in sweetness."

On the basis of all the evidence Lamb concluded that English growing-season temperatures during the Mediaeval Warm Period must have averaged 0.7–1.0°C higher than those of the early to mid 20th century, and on the Continent 1.0–1.4°C higher. The evidence also suggested that annual rainfall was higher than now but the summers drier: i.e. there was greater similarity to a Mediterranean-type climate. (This is consistent with what computer models project for climate warming in the region.)

Some researchers have since contested Lamb's conclusions and those of others working from historical, geographical and geomorphological evidence. Criticism has come particularly from those researching the green-house effect using computer models. Their 'consensus' view, reflected in the various (e.g. 2007) publications of the officially constituted Intergovernmental Panel on Climate Change (IPCC), is that the warmth and warming of the late 20th century are unprecedented in the last millennium and probably longer. Justifying this position logically necessitates being able to disprove the existence of greater warmth in mediaeval times.

A representative and detailed critique from the 'consensus' perspective is that of P.D. Jones and Mann (2004), which seeks generally to refute the older evidence. (Note that several researchers named Jones feature in greenhouse and related literature. To avoid confusion I specify their initials throughout.)

While some of the criticisms and reservations are valid, those on viticulture reveal serious misconceptions which we are here in a position to correct. The relevant passage in their paper is as follows.

> Monks in mediaeval England grew vines as wine was required for sacramental purposes. With careful husbandry, vines can be grown today, and indeed, vineyards are found as far north as southern Yorkshire. There are a considerably greater number of active vineyards in England and Wales today (roughly 350) than recorded in the *Domesday Book* of AD 1086. Vine growing persisted in England throughout the millennium. The process of making sparkling wine was developed in London (by Christopher Merret) in the 17th century, fully 30 years before it began in the Champagne region of France. Thus the oft cited example of past vine growing in England reflects little, if any, on the relative climate changes in the region since mediaeval times.

Three comments can be made on this passage. First, it conveys an impression that English mediaeval viticulture was confined to monasteries making sacramental wine: a requirement that could be met from walled or otherwise highly sheltered small plantings. In fact it appears from the researches of Lamb and Unwin, already cited, that post-Conquest viticulture was mostly in the open and part of the normal secular economy. Significant numbers of vineyards were over 10 acres or in operation for over 100 years.

Second, Jones and Mann fail to account for differences in the grape varieties grown. Those at the cold limit at any time are of necessity the earliest-maturing available. Conservatively assuming no genetic progress in earliness between the mediaeval varieties and the traditional varieties grown in later centuries, it is reasonable to infer that the English mediaeval vines belonged typically in maturity groups 3-4 as now classified (see Chapter 10, Table 10.1). They probably included varieties such as Pinot Noir and Elbling (both Maturity Group 3) and Aligoté (Group 4) which were the archetypal vines of northern France and Germany through the Middle Ages and beyond (see Robinson 1986).

But the varieties grown now in England are overwhelmingly the products of German and other breeding for earliness that started with the climatic cooling of the late 19th century. These new varieties fall mainly into maturity groups 1-2. Using the methods described in Chapter 11, we can estimate that the earliest of the older varieties would have needed growing-season temperatures not less than 0.5–0.6°C above those of the late 20th century to reach comparable maturity. Rather than refuting Lamb's temperature estimate for mediaeval England, the viticultural facts support it.

Jones and Mann's third misconception concerns the history of 'champagne' sparkling wine. It is true that the process for its making was first developed in London, in the 1660s. What made this possible was the development by London manufacturers of very strong glass bottles, together with the use of corks tied down with string as had been used for ale etc. since Elizabethan times. The base wines were imported in cask from Champagne, but it was in London that the practice grew of adding a *dosage* of sugar or molasses at bottling to promote secondary fermentation in bottle.

Pioneer of importing the still Champagne wine to London was a Frenchman, Seigneur Charles de Marguetel de Saint Denis de Saint Evremond (1613–1703), who had fled to England as a political refugee in 1661 and became a favourite (along with champagne) at the court of King Charles II. The king rewarded him with a lucrative sinecure as 'Governor of the Duck Islands', the latter being a group of small islands in the St James Park canal used for decoying wild duck. It was not until late in the century that the necessary bottles and corks became available in France and the monk Dom Perignon was able to perfect the champagne-making process. André Simon (Simon 1971) tells the full story.

Clearly the history of champagne had no connection at all to English viticulture, which by the 15th century had become essentially defunct following successive climatic crises, increased competition from Bordeaux after its accession to England with Eleanor of Aquitaine in 1152, and labour shortages following the Black Death. Barton (1991) cites records showing a small recovery in the 16th and 17th centuries, but it has never been suggested that any wines produced then were significant in national commerce. A larger recovery reportedly followed in the relatively warm summers of the 18th century, but these vines again disappeared in the 19th century, probably during its cold early decades. Modern English viticulture only started after World War II.

Changes in European viticultural climates became better documented after the mid 14th century through detailed records of officially proclaimed dates for the start of harvest (taking into account the 10-day calendar advance with the switch on the Continent from Julian to Gregorian calendars in 1582). Combined with parallel studies of tree-rings, this has made possible a fair estimate of changes over time in both decadally averaged maturity dates and seasonal average temperatures. Figure 12.2 shows generalized maturity dates for western Europe from 1370 to 1870 after Pfister (1988), based largely on the work of Le Roy Ladurie and Baulant (1980).

As Le Roy Ladurie (1971) has emphasized, recorded maturity dates in the

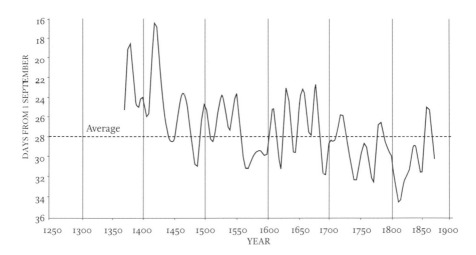

Figure 12.2. Estimated average dates of starting vintage in Western Europe, AD 1370–1880. From Pfister (1988).

most recent centuries must allow for changing ideals of wine style and the technical means for their production. Mediaeval red wines in western Europe appear mostly to have been light and perhaps closer to rosés than to modern red wines. The change to fuller-bodied styles from the 18th century onwards would have followed the growing practice of laying down superior red wines for maturation in bottle. Then in the mid–late 20th century the rise of more efficient fungicides, that made full grape ripening safer, together with a market taste for riper-style wines have undoubtedly carried the trend forwards. Despite this, Figure 12.2 still shows good general agreement with the historical and other non-viticultural indicators of climate over the centuries.

By the start of the graphs, European average temperatures had already fallen from their heights of the Mediaeval Warm Period, most notably in the calamitous runs of cold and wet seasons in the early and mid 14th century that brought about widespread famine and presaged the outbreak of the Black Death (1348–50).

The late 14th century saw a brief return of notable warmth, but there followed an even steeper temperature decline towards the following mid-century and further decline late in that century. This appears to have been when much of the old northern viticulture was finally abandoned, including that of the Paris region which had previously enjoyed a large export trade. There was widespread abandonment of marginal farms and villages generally in northern Europe, continuing the trend started in the previous century (Lamb 1977, 1984).

The 15th century is generally taken as the start of the so called Little Ice Age, a period of intermittently low temperatures and advancing glaciers, from which the world did not start to emerge until the mid 19th century or even later. Sea levels (Figure 12.1) suggest the 15th century to have been particularly cold worldwide.

One characteristic of the Little Ice Age that seems generally agreed on is that the main chilling was in winter (P.D. Jones et al. 2003; Luterbacher et al. 2004). Periodic extremes of winter cold resulted from persisting anticyclones over the North Atlantic to Scandinavia associated with negative values of the North Atlantic Oscillation Index, or else from westward extension of the normal winter anticyclone over Siberia. Both these sets of synoptic conditions directed polar air southwards or westwards over western Europe.

A second characteristic of the Little Ice Age was its extremely intermittent nature, with intervening periods having quite warm summers. For instance Figure 12.2 shows European vine growing seasons equal to or warmer than the 20th century average towards the middle of the 16th century, and especially through much of the 17th century. The latter fits known wine history. Champagne in the 17th century was famed for its still red wines made from Pinot Noir, which were said to rival those of Burgundy, if a little lighter (Simon 1971). Few Champagne red wines of the 20th century can match that.

The precipitous renewed temperature fall in the last decade of the 17th century caused much agricultural stress throughout Europe. I have speculated elsewhere (Gladstones 1992) that it contributed to the rise of sparkling champagne in London and in France: in many seasons the grapes did not ripen enough to make saleable dry still wines.

Temperatures started to rise again early in the 18th century, but not until after unprecedented cold in the 1708/09 winter caused widespread vine killing through much of the continent, even in southern France. This led to the final wave of abandonment of many northern vineyards, and greatly increased planting further south in Bordeaux and on the French south coast. Improving transport from then on also contributed to the growing competitiveness of the south, and to the flood of planting on the southern coastal plains to supply *vin ordinaire* to the burgeoning industrial markets of the north.

Thus the history of European viticulture bespeaks a climatic cooling and southerly migration of vine plantings from the Mediaeval Warm Period (about AD 900–1300) down at least to the 19th century. Bradley (1991) has discussed this is some detail, suggesting an overall fall in European growing season temperature of 0.75°C compared with the warmest historical decades.

Some modern researches on vintages tend to corroborate the long-term European temperature decline, and to suggest that it continued until quite recently.

It is a general (and reasonable) belief that the dominant grape varieties of long-established European viticultural regions are those that have proved over the centuries to have the best-fitting phenologies. Yet we find that even in the supposedly very warm late 20th century, most of the best vintages have been in the warmer and drier seasons that lead to early vintage. Two studies demonstrating this for Bordeaux have been those of Ashenfelter et al. (1995) and G.V. Jones and Davis (2000a, 2000b).

Ashenfelter et al. used an index of the prices that Bordeaux vintages from selected chateaux subsequently achieved on the London auction market as mature wines. For vintages from 1952 to 1980 they found that the highest-valued typically came from warmer-than-average growing seasons, with dry ripening periods and most rainfall in the preceding winter. Interestingly, seasonal evaluation in these terms gave far better predictions of ultimate value than did market evaluation of the young wines.

G.V. Jones and Davis (2000b) used records collected by the University of Bordeaux of vine phenology, mature berry size and composition, and vintage rating for 10–15 leading chateaux, matched against weather data from the Bordeaux Mérignac Airport from 1949 to 1997. They likewise showed that the warmer seasons and earlier vintages gave on average the best wines, even late in the century. Vintage quality for both Cabernet Sauvignon and Merlot grape varieties was closely and positively correlated with berry sugar content at harvest, and negatively with acid content.

Both studies seem to indicate that through much of the 20th century these historically adapted varieties ripened less fully than ideal and than they had once done. Consistent with this has been the substantial 20th century disappearance of the late-maturing and hard-to-ripen variety Petit Verdot, which was once important and esteemed in Bordeaux (Dubourdieu 1990). The same perhaps applies to Carmenère, another esteemed but late-maturing variety that was reportedly prominent in 18th century Bordeaux vineyards, but has since largely disappeared from there (Daurel 1892, cited by Robinson 2006).

Grifoni et al. (2006) reported a comparable study that matched the vintage ratings of five leading central and northern Italian wine regions against measures of mean temperature and rainfall over the years 1970–2002. Again there was a clear indication that the warmer and drier growing seasons and ripening periods gave the best vintages. Growing-season average mean temperatures in this case covered the range 18.3–20.5°C. G.V. Jones et al. (2005) present similar data and conclusions for a variety of regions in Europe and elsewhere.

Such findings are significant for future European viticulture, since the currently superior seasons correspond to what computer models suggest may more regularly occur under future moderate global warming. The findings also confirm that late 20th century warmth may approach, but almost certainly does not equal, that of the Mediaeval Warm Period.

Most evidence cited so far, other than for sea levels, comes from Europe. The question then arises: how indicative is European climate of that experienced elsewhere?

12.1.3 Were the climate changes global?

The extent to which European climatic fluctuations have reflected similar and synchronous fluctuations world-wide has always been controversial. The 'consensus' view, as exemplified by P.D. Jones and Mann (2004), and Osborn and Briffa (2004), is that they largely do not. This view emphasizes apparent differences in timing, location, type and extent of climate change within both the Mediaeval Warm Period and the Little Ice Age, while acknowledging their existence in broad terms.

It is true that climatic changes, by their nature, cannot be fully identical or synchronous across environments. We know that the interiors of continents heat and cool faster than coasts and especially oceans. Regions influenced by the incidence of sea ice or winter snow cover – for instance Siberia – will undergo still greater and faster changes, since the advance of either accelerates local cooling through reflecting solar radiation away, while their retreat conversely accelerates local warming. This is a case of positive feed-back under both warming and cooling. Thus high latitudes of the Northern Hemisphere undergo wider and faster climatic fluctuations than the tropics or the ocean-buffered Southern Hemisphere.

Further, any induced changes in the major ocean currents can impinge differentially on the coasts and hinterlands they influence. The massive inertia of the currents also means that resulting climate changes can persist well after their original impetus has ceased.

These reservations accepted, the available evidence does suggest that the major (e.g. centennial) temperature fluctuations of the historical period have been essentially synchronous across the continents and hemispheres. The evidence comes not only from sea levels, as already described, but also from studies of glaciers (Röthlisberger 1986, cited by Wigley and Kelly 1990;

Oerlemans 2005); oxygen isotopes in ice cores (Thompson et al. 1986); and on more detailed time scales from tree-ring and other chronologies (Esper et al. 2002; Cook et al. 2002; Yang et al. 2002).

It is of interest that the Mediaeval Warm Period as delineated in the above studies coincided with often long periods of extreme drought in the Great Plains and California regions of the USA, far exceeding the 'dust bowl' of the 1930s (Stine 1994; Laird et al. 1996; Woodhouse and Overpeck 1998; Cook et al. 2004). A generally considered reason was poleward displacement of the weather systems during this warm period, leading to less winter rain in California, and more summer settling of high pressure cells over the plains east of the Rockies that blocked summer rain-bearing winds from the east and south. Similarly Lamb (1977) noted a high incidence of drought from central Europe to central Asia over this period, with low levels of the Caspian Sea. Parallel to the North American droughts, this started to reverse early in the 13th century. During the cooling of the next 200–300 years the Caspian rose to eight metres above its 20th century level.

The droughts of the mid-latitude continental interiors during the Mediaeval Warm Period agree with model projections for global warming. Warming may increase evaporation and hence rainfall at these latitudes in ocean-influenced climates, where ample water or vegetation surface is available for evaporation. But beyond the reach of this moisture, e.g. in the rain shadow of the Rockies, or anywhere inland where atmospheric high pressure cells predominate in summer, the effect can be progressive soil and atmospheric drying and thence escalating, self-sustaining drought. The history and prospect of continental interior droughts holds an important message for viticultural (and other) planning, should warming continue for whatever reason.

12.2   Natural causes of climate change

12.2.1 *Earth-sun geometry*
The combined roles of changes in the earth's elliptical orbit around the sun in a 100,000-year cycle, of pendulum-like swings in the tilt of its axis of revolution over a 40,000-year cycle, and of a circular wobble in its axis of revolution, in a 22,000-year cycle are now recognized as controlling multi-millennial changes in the earth's climate such as the coming and going of the ice ages over the past million or so years. Together these form a highly predictive model of climate changes across that timescale, now known as the Milankovich Model after its proponent Milutin Milankovich.

Fairbridge (1962) has dealt in some detail with the climate and sea-level changes under this model since the end of the last ice age about 10,000 years BP (before present). By that time the Milankovich annual and (especially) summer radiation for the Northern Hemisphere had reached its inter-glacial peak, but it was not until about 6,000 years BP that mid-latitude temperatures and sea levels had caught up owing to the lag in melting of the polar ice sheets. Estimated sea levels and global average temperatures were then 3–4 metres and 2–3°C above those of the 20th century.

Milankovich solar radiation for the Northern Hemisphere, which is critical for global temperatures because of the high-latitude Northern Hemisphere's particular sensitivity to radiation changes, has since fallen gradually and been accompanied by an irregular fall in both sea level and temperature. The period from 6,000 to 4,000 BP saw wide fluctuations in sea levels and, presumably, temperature, the former ranging some two to three metres above and below present levels (Fairbridge 1962). The fluctuations have since become progressively less. The last complete minor peak was that of the Mediaeval Warm Period, while the late 20th century appears to represent a comparable one.

From the Mediaeval Warm Period to the 19th century the overall trend was downwards, with the analysis of Bradley (2000) suggesting an underlying fall of about 0.2°C that might be attributed to Milankovich factors. It is the background against which we can now turn to reasons for the natural climatic fluctuations over shorter (annual to centennial) timescales that have been documented within the last millennium.

12.2.2 Volcanic activity

The temporary cooling effect of stratospheric dust veils thrown up by major volcanic eruptions has been long known. Credible measurements of the temperature responses, their extent and duration, and long-term effects on climate have only come recently, however. Two authoritative studies are those of Briffa et al. (1998a) and Shindell et al. (2003).

Primary information comes from worldwide temperature measurements following well-documented eruptions in the modern era, which can then be extrapolated back to earlier historically recorded eruptions. In most cases these can be corroborated, or unrecorded eruptions inferred, both as to date and intensity by the study of sulphate deposition in dateable polar ice cores.

Briffa et al. related tree ring densities and associated estimates of Northern Hemisphere summer temperature anomalies to the volcanic record back to AD 1400. Some relationships are apparent that accord with other records. For instance the coldest decades in the mid and late 15th century, the late 16th century and the mid 17th century all coincided with clusters of major eruptions. No clear evidence connects the extreme cold of the 1690s to volcanic activity, but the Tambora eruption of 1815, the strongest and only one in historical times to be rated the maximum Volcanic Explosivity Index (VEI) of 7, appears unequivocally to have contributed to the European very cold seasons of 1816 and to a lesser extent 1817–18. The estimated Northern Hemisphere summer temperature depression for 1816 was 0.5°C. A marked cooling apparent in the modern temperature record between 1880 and 1915 (see Section 12.5) may have been partly due to a succession of VEI 5 and 6 eruptions, while the subsequent rapid warming of the 1920s has sometimes been attributed to a contrasting lack of them.

Finally the Pinatubo eruption of 1991 (VEI 6) brought about a Northern Hemisphere cooling in summer and autumn of the following year in the order of 0.4–0.5°C (P.D. Jones 1994). However no such cooling was noted in winter-spring. This accorded with the more general observation of Robock and Mao (1992) that volcanic dust veils can actually cause winter warming in some high-latitude regions, a point that Shindell et al. (2003) also emphasize. This has general significance, in that the most widely used proxy measures of past volcanic effects, tree-ring densities and widths, reflect primarily summer conditions. They therefore probably over-estimate effects on climate overall. Nor, in the case of Pinatubo, did Jones' figures show measurable cooling in the Southern Hemisphere, reducing its significance for global climate still further.

Shindell et al. examined mainly the longer-term (decadal to centennial) impacts of vulcanism. They concluded that once yearly temperatures are smoothed to decadal or longer averages, the effect is not statistically distinguishable from background variability. By contrast the contribution of solar phenomena, which constitute the only known other natural element capable of short to medium-term influence over climate, is highly significant and largely consistent across seasons and regions.

It follows that an understanding of solar phenomena is critical, not only to understanding natural climates, but also to assessing any recent and future complementary role of anthropogenic greenhouse gases in climate change. However we will defer considering the latter aspect to Sections 12.3 and 12.4, and concentrate for now on the former.

12.2.3 Solar irradiance and magnetic field

Subsection 12.2.1 has already touched on the mutual relationships of sea levels, sunspot numbers and imputed temperature over the last millennium, as highlighted by Fairbridge (1962). Since then, Eddy (1976, 1977) and Stuiver and Quay (1980) have developed further the relationships among climatic cycles and those of sunspots and tree-ring ^{14}C: the last as a proxy of the sun's variable strength of magnetic field.

The relevance of ^{14}C is that it is produced in the atmosphere from normal ^{12}C under bombardment from incoming cosmic radiation in proportion to radiation intensity. The magnetic fields of both the earth (fluctuating slowly on a multi-millennial timescale) and of the sun (varying on much shorter timescales) act as shields that divert cosmic radiation away from earth. Thus ^{14}C incorporated into tree rings relates inversely with current solar activity and strength of magnetic field. Low ^{14}C in tree rings indicates high solar activity, magnetic field strength and temperature at their times of deposition; high ^{14}C the reverse. Note, however, that this relationship breaks down after about AD 1850, owing to increasing contamination by atmospheric CO_2 produced from burning fossil fuels (the 'Suess Effect').

The beryllium isotope ^{10}Be provides another inverse indicator of solar magnetic field strength, being formed in the upper atmosphere when cosmic rays interact with nitrogen and oxygen nuclei. It is carried down and accumulates in polar icecaps with successive, and dateable, winter snowfalls. The particular utility of ^{10}Be in ice cores is that it can trace variations in the solar magnetic field back over many millennia.

A final proxy to be noted is the incidence of (Northern Hemisphere) mid-latitude aurora sightings, which Silverman (1992) discusses comprehensively. These are directly related to solar magnetic phenomena. Over the 500 years or so for which records are available they essentially parallel the sunspot data, though with smaller amplitudes of their approximately 11-year short-term cycles.

The following synthesis comes from the publications of Eddy (1977), Stuiver and Quay (1980) and Silverman (1992), using records of ^{14}C, sunspots, auroras and the magnetic index Aa. (The last measures solar-caused disturbances in the earth's magnetic field, for which records go back 200 years. It can also be inferred for earlier periods from the ^{14}C record. Stuiver and Quay give details.)

Together these proxies provide a body of evidence that is remarkable for its consistency, both among proxies and with the historical record as described in Section 12.1. The high solar activity that persisted through the Mediaeval Warm Period diminished rapidly after about AD 1270, reaching a minimum between 1290 and 1340 that is known as the Wolf Minimum

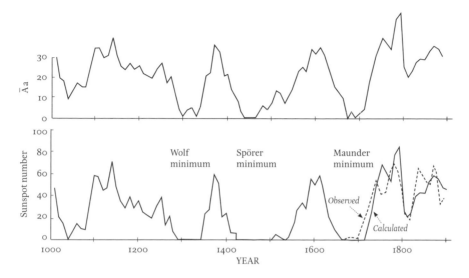

Figure 12.3. Estimated geomagnetic Aa indices based on ¹⁴C data relative to sunspot frequencies for AD 1000–1900. From Stuiver and Quay (1980).

(Figure 12.3). Renewed solar activity and warmth between then and a little after 1400 gave way to the Spörer Minimum, which lasted somewhat indefinitely into the early 1500s before a fairly prolonged recovery of solar activity and relative warmth to the mid to late 1600s. Precipitous falls late in that century signalled the Maunder Minimum of about 1670 to 1710: a time when vastly improved telescopes and extensive observations proved beyond doubt a more or less complete absence of sunspots and Northern Hemisphere mid-latitude auroras.

The Maunder Minimum is generally taken as a turning point, after which the trend of solar activity and imputed earth temperatures has been irregularly upwards and fairly well documented. Some key references for this latter period are those of Hoyt and Schatten (1993), Lean et al. (1995), Lean (2000), Solanki and Fligge (1999), Solanki et al. (2000), Lockwood and Stamper (1999) and Lockwood (2001). Lean and Rind (1998) and Robertson et al. (2001) give good general reviews.

Since the beginning of direct solar irradiation measurements in 1980, proxy measurements of past solar activity have increasingly been translated into irradiance estimates. Figure 12.4, from Hoyt and Schatten (1993), shows one such reconstruction from 1700 to the late 20th century.

Irradiance trends so estimated are clearly upwards from 1700, but still with marked fluctuations. High solar activity through much of the mid to late 18th century preceded the deep, but short, Dalton Minimum of about

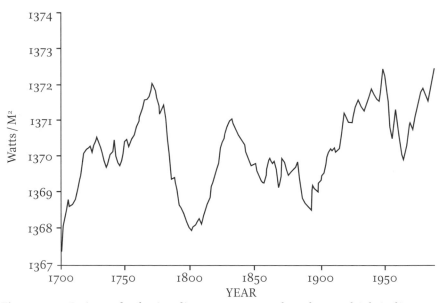

Figure 12.4. Estimated solar irradiance 1700–1992, based on multiple indices. After Hoyt and Schatten (1993).

1800–1830 ('the years of the Dickensian winters'). There followed a further moderate irradiance peak in the mid 19th century. The auroral record (Silverman 1992) differs a little in showing particularly marked magnetic activity between 1860 and 1880. The great pre-phylloxera Bordeaux vintages, with a succession of very early vintages, came especially in the 1860s and early 1870s (Ray 1968; Gladstones 1992). I underline the point because the proxy evidence is at odds with generally accepted interpretations of the thermometer record for the time, as will be discussed in Sections 12.5 and 13.2.

The most prominent recent feature of the solar activity proxies and estimated solar irradiance is their steep and largely continuous increase throughout the 20th century, following a moderate dip around the start of the century (Figure 12.4). Lockwood (2001) estimated solar magnetic flux to have increased 140% since 1900 and 34% since 1963, with closely correlated falls in the cosmic rays reaching earth. These figures point to a marked natural warming through the century.

Some uncertainty remains about proxies for the latter part of the century. Sunspot numbers stabilized after about 1960, but peak-to-peak length of the '11-year' sunspot cycle continued to shorten. There is now evidence (Friis-Christensen and Lassen 1991; Lassen and Friis-Christensen 1995; Lockwood 2001) that this is a better indicator of growing solar activity than simple sunspot numbers. Jirikowic and Damon (1995) and Solanki et al. (2004)

suggest that solar activity has now approached or reached its level of the Mediaeval Warm Period.

Evidence from earlier millennia within the present inter-glacial, derived from correlations with ^{10}Be and ^{14}C records, likewise indicate climatic control by solar activity overlying the gradual long-term changes due to earth-sun geometry. Evidence comes, for instance, from latitudinal fluctuations in the deposition of detritus from sea ice (Bond et al. 2001); cyclic variation in deposition of Alaskan lake sediments (Hu et al. 2003), and in the advance and retreat of Alaskan glaciers (Wiles et al. 2004); and from the nature of planktonic sediments in the Gulf of Mexico (Poore et al. 2004). All bespeak quasi-cyclic latitudinal migrations of the climate belts in synchrony with solar changes. Shindell et al. (2001) argued similarly for European climates through solar-controlled shifts in the North Atlantic Oscillation.

All of these echo in miniature the migrations of the climate belts that Fairbridge (1962) postulated for the coming and going of the ice ages. Warming by increased solar activity or influence brings about poleward migration of the Inter-Tropical Convergence Zone and its associated summer monsoonal rains, with contraction of the polar air masses and poleward migration of the rain-bearing mid-latitude storm belts. Cooling brings about the opposite migrations.

12.2.4 *Modelling pre-industrial temperatures*

We have previously noted the scepticism of climate change modellers concerning historical and proxy indicators of past temperatures. This school holds that the resulting estimates cannot be taken seriously unless corroborated by direct matching against modern measured temperatures. In the words of P.D. Jones and Mann (2004): "The use of instrumental climate data records is an essential component of high-resolution paleoclimatology, as it provides the quantitative information against which the proxy climatic indicators must be calibrated." Similarly Mann et al. (1998) wrote:

> One can think of the instrumental patterns as 'training' temperatures against which we calibrate or 'train' the much longer proxy data (that is, the 'trainee' data) during the shorter calibration period in which they overlap. This calibration allows us to subsequently solve an 'inverse problem' whereby best estimates of surface temperature patterns are deduced back in time before the calibration period, from the multiproxy network alone.

Most attempts to do this have used either or both of tree-ring thickness or density, and deep bore-hole temperature measurements.

Such methods have problems of their own, as their protagonists generally concede. For instance tree-ring thickness and density only reflect temperature in a few high-latitude or high-altitude environments where temperatures are regularly low enough to limit growth directly through the summer growing season. Temperatures rising above the response range do not register, while extraneous factors such as drought, disease, pest attack and fire may also periodically limit growth. Further, rising atmospheric CO_2 since the start of the industrial revolution can be expected to produce a plant growth response of its own, as La Marche et al. (1984) discuss. This covers more or less exactly the period of calibration against temperature records, but has otherwise been little acknowledged in the literature. Finally, Briffa et al. (1998b, 1998c, 2001) report a widespread failure of tree-ring densities to match recorded temperature rises in the final decades of the 20th century; indeed, densities fell while recorded temperatures rose steeply, a phenomenon for which the authors could offer no explanation.

Unsurprisingly the backward projections of temperature based on tree-ring width or density have given mixed results, which can depend as well on the mathematical methods used. The projections of Briffa et al. (1990, 2001) suggested mostly low pre-industrial temperatures, with some indication of the Little Ice Age but little of a Mediaeval Warm Period. On the other hand Esper et al. (2002), using a different reconstruction approach, reproduced approximately the patterns of other proxy data, if with some displacement in time.

Use of borehole temperatures poses even greater problems. It depends on the slow diffusion of surface heat to depth, so that transient (e.g. decadal to centennial) changes in surface temperature are reflected in anomalies in the trend of temperature with depth. Deepest anomalies are the oldest. However they can be confounded with many other factors, such as the differing heat conductivities of rock materials (both vertically and laterally), surface topography and vegetation cover, and water movement at depth. Nevertheless several studies of the combined results from many boreholes concur in finding that Northern Hemisphere land surface temperatures have risen about 0.8–1.0°C from the beginning of industrial times to the late 20th century, in agreement with surface thermometer measurements, although Mann et al. (2003) re-work previously published data to argue a smaller rise. For examples of these studies, and detailed discussion of their pitfalls, see Shen et al. (1995); Pollack et al. (1998); Huang et al. (2000); Harris and Chapman (2001); Mann et al. (2003).

Finally, and most controversially, we come to a series of studies at the

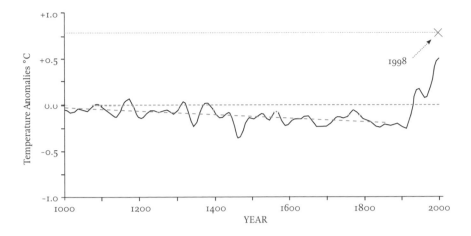

Figure 12.5. The 'hockey stick' graph of estimated Northern Hemisphere temperature anomalies from the 1902–1980 average mean, AD 1000–2000. From Bradley (2000), after Mann et al. (1999).

heart of the climate change debate, that use multiple proxies to reconstruct pre-industrial temperatures by regressing from the template of the thermometer period. Mann et al. (1998, and particularly 1999) concluded in this way that pre-industrial temperatures fluctuated only a little around a millennial slow decline of about 0.2°C up to AD 1900; by contrast, subsequent recorded temperatures to the end of the 20th century rose steeply by some 0.8°C. The resulting graph (Mann et al. 1999; Bradley 2000) became known from its shape as 'the hockey stick' (Figure 12.5) and was widely publicized in IPCC and other literature. Its implication was clear. The decadal to centennial impact of natural forces on climate is very limited, beyond minor short-term volcanic effects and some internally generated chaotic behaviour. Unprecedented warming in the 20th century must therefore be man-caused.

Later review (von Storch et al. 2004; Moberg et al. 2005; see also Osborn and Briffa 2004) showed, however, that the imputed lack of pre-industrial climate variability resulted chiefly from the mathematical treatments used. Their regression-based methods, in attempting to remove statistical noise, dampened the amplitude of short to medium-term climatic variations by a factor of two or more. Correcting for this error brings model temperature estimates for the past millennium closer to consistency with the various historical and proxy data: as do new regression methods that preserve multi-decadal and centennial variance (Moberg et al. 2005; Hegerl et al. 2007).

The final step in modelling climatic changes and attributing their causes must incorporate the contribution (if any) from rise of anthropogenic green-house gases and pollutants in the atmosphere since the beginning of the industrial revolution.

12.3   Anthropogenic causes of climate change

It is now widely believed that the increasing addition to the atmosphere of combustion by-products from fossil fuels used in industrial and other human activities can influence, perhaps dominate, global climate change in the direction of warming.

Chief among them is carbon dioxide (CO_2). This assumes primary importance because of its rapid increase with growing worldwide indus-trialization. (Water vapour plays a much larger greenhouse role, but is not of direct anthropogenic origin. Its amount in the atmosphere varies with temperature changes whatever their cause.) Other greenhouse gases will be mentioned only briefly; except where necessary we will here adopt the usual approach of considering them as part of an overall 'CO_2 equivalent'.

The warming effect of greenhouse gases stems from the fact that they are transparent to most incoming short-wave solar radiation, but to varying degrees opaque to outgoing long-wave radiation from the relatively cool earth. Outgoing heat is absorbed and trapped in this way, mainly in the lower atmosphere, or troposphere; it warms that air layer, which in turn radi-ates some of the heat back to lower air and the earth's surface. As greenhouse gases increase, so must tropospheric and surface temperature rise until the resulting greater outgoing radiation establishes a new equilibrium. However these changes and the final equilibrium reached can be much influenced by positive and negative feedback processes, as will be discussed in Section 12.4. For a comprehensive account of all greenhouse gases and their effects, see Shine et al. (1990).

12.3.1 *Carbon dioxide and water vapour*

The difference between the earth's average surface air temperature (288°K, or 15°C) and its effective heat emitting temperature as seen from space (255°K, or minus 18°C) is a measure of its total existing greenhouse effect, which is due principally to water vapour and (to a smaller extent) CO_2. The proportion due to CO_2 is estimated at about 25% (Shine et al. 1990). These authors, Kiehl and Ramanathan (1983), Ramanathan et al. (1987) and Ramanathan (1998) give accounts of the physics involved.

Main interest lies in the marginal effects of additional CO_2, which has already risen from its pre-industrial level of about 280 parts per million (ppm) to 385 ppm in 2008. Its greenhouse effect is, however, not directly proportional to concentration. Pre-industrial levels already saturated their greenhouse function almost completely, so that further increments have steeply diminishing effects. These are now accepted as being proportional to the base 10 logarithm of CO_2 concentration. That is, temperature rise would be the same for each successive doubling of CO_2.

Efforts to model temperature responses to past and potential rises in atmospheric CO_2 have a long scientific history, but those incorporated into credibly realistic climate systems are generally taken to date from the work of Manabe and Wetherald (1967, 1975). They estimated that a doubling of atmospheric CO_2, once its effects had come into equilibrium, would directly raise global average temperature by 1.3°C. The resulting enhanced evaporation and rise in atmospheric water vapour would raise it a further 1.0°C, a total of 2.3°C. However they cautioned that this assumed constant cloud cover: any variation in cloudiness could markedly affect the result.

Rasool and Schneider (1971) estimated a much smaller direct temperature rise from doubled CO_2 of 0.6°C, with a further 0.2°C from extra water vapour. Several other authors around the same time produced still lower figures. But later review (see Schneider 1975) suggested a variety of short-comings in their calculations or assumptions. Most estimates since have settled around a direct CO_2 (or CO_2-equivalent) doubling effect around 1.1–1.2°C, e.g. Hansen et al. (1981); Lal and Ramanathan (1984); Schlesinger (1985); Ramanathan et al. (1987). With allowance for resulting extra water vapour, but no other feedbacks, this gives a warming from doubled CO_2 that is generally accepted among modellers of a little over 2.0°C. This in turn equates to an enhanced 'radiative forcing' at the earth's surface of about 4.0 watts per square metre (Wm^{-2}) (Hansen et al. 1981; Ramanathan et al. 1987).

12.3.2 Anthropogenic aerosols and other pollutants

Whereas the simple effects of greenhouse gases and water vapour on surface air temperatures are generally thought of as well understood, those of other emissions from burning fossil (and other) fuels are far from being so. Chief among them is sulphur dioxide (SO_2), which forms sulphate aerosols that can persist in the lower atmosphere for up to a week of two. These produce cooling by scattering and reflecting away incoming sunlight, in the same way as stratospheric volcanic aerosols. It is also thought that as supplementary cloud condensation nuclei, they produce increased numbers of consequently lighter droplets, deferring their coalescence and fall as raindrops. This both prolongs the lifespan of clouds and makes them more reflective.

The idea that some other anthropogenic factor might be counteracting the warming due to CO_2, water vapour and other greenhouse gases came with a realization by researchers that the measured warming since the start of thermometer records was less than computer models predicted they should be. Some key papers developing and modifying the idea are those of Wigley (1989), Charlson et al. (1991, 1992), Kiehl and Briegleb (1993), Mitchell et al. (1995), Mitchell and Johns (1997) and Hansen et al. (1997). Charlson et al. (1992) estimated from physical principles that present sulphate aerosols could result in a globally averaged cooling that would at least substantially offset the warming caused by greenhouse gases. However, the later estimates were more conservative.

Identifying the real-life temperature effects of anthropogenic aerosols should in theory be simple, since their main production is concentrated in industrial areas, while their short lifespan means that action must mostly be within those areas or fairly close down-wind. Yet the literature yields surprisingly few studies on that point. The most relevant I have been able to find is that of Engardt and Rodhe (1993), which gave mixed results. They identified major sources of SO_2 emissions in central and eastern Europe and, to lesser degrees, eastern USA and eastern China. All three centres showed evidence of minor summer cooling over time (0.2–0.4°C) relative to the rest of the Northern Hemisphere, but in eastern China and eastern USA there was a greater (0.5–0.9°C) relative warming in winter. The south-east of the USA cooled slightly in both winter and summer. Complicating the issue is that some of these changes might equally be ascribed to evolving patterns of land use, as will be discussed in subsection 12.3.3; but in any case there was no indication of major overall cooling.

The second line of evidence comes from comparing temperature trends in the two hemispheres. Nearly all anthropogenic SO_2 emissions have historically been in the Northern Hemisphere, and the short life of their products in the air largely precludes inter-hemispheric exchange. As Schwartz (1988) has shown, late 20th century sulphate concentrations in Northern Hemisphere air are 3–4 times greater than in the Southern Hemisphere. This should have resulted, if sulphate aerosols have a significant overall cooling effect, in a clear inter-hemispheric divergence of temperature trends over the 20th century. The record shows no such divergence: see for instance Wigley (1989); P.D. Jones et al. (1999); P.D. Jones and Moberg (2003). Likewise Schwartz (1988) detected no relative enhancement of Northern Hemisphere cloud reflectivity, as Charlson et al. (1991, 1992) and others suggested should have occurred. Nor did Norris (1999) find any greater increase in ocean cloud cover in the northern mid-latitudes as compared with the tropics and Southern Hemisphere, as might be expected if anthropogenic aerosols increased cloud formation other than locally.

A possible contributing reason for the lack of any evident worldwide cooling from anthropogenic aerosols is that other components may counter-balance those derived from SO_2. Kaufman and Koren (2006 and references therein) showed that the effects of smoke from burning biomass in the tropics tend if anything towards warming. This is because the dark-coloured smoke aerosols absorb heat in the cloud zone, potentially reducing cloud.

Finally, the role of tropospheric ozone pollution has to be considered. This occurs both from industrial pollution and, mainly in the tropics, from biomass burning (Kiehl et al. 1999). Ozone acts as a greenhouse gas in the troposphere by absorbing both shortwave and longwave radiations. As with aerosols the effects are localized because of ozone's limited lifespan, and are also seasonal because the lifespan is temperature-dependent, being greatest in summer. Kiehl et al. estimate that global positive forcing by tropospheric ozone in summer approximately halves what would otherwise be a direct negative forcing by sulphate aerosols, from -0.76 Wm^{-2} to -0.40 Wm^{-2}.

A best overall conclusion seems to be that anthropogenic aerosols and ozone pollution, between them, can offset global greenhouse and natural warming by at most only a very small amount. They cannot account for the discrepancy between modelled and measured recent warming.

12.3.3 Land use effects

Changes in land use have hitherto featured little in global climatic studies, but evidence is now available that they can have significant impacts locally, regionally and perhaps globally. Either warming or cooling can result.

One of the best-studied cases is that of the midwestern and eastern USA, as documented by Bonan (1997, 2001). Since the mid 19th century the Midwest has experienced almost total replacement of woodlands and prairie by cropland. Over that time measured mean temperatures have fallen and diurnal temperature ranges narrowed, a process that has continued well into the 20th century (Karl et al. 1984). Proffered explanations are two-fold. First, crops with their shiny leaves reflect away more light energy than the native vegetation. Second, their large biomasses and leaf areas in summer-early autumn cool the adjacent atmosphere by transpiring large amounts of water in this well-watered region. (Winter transpiration is small whatever the vegetation.) The cooling is mostly in the daytime, hence a reduced diurnal range. In recent decades the bigger crops obtained through intensive fertilizer use, and in some cases irrigation, have doubtless enhanced the process further even after crop area has stabilized.

The opposite happens when tropical rainforests are cleared, as Couzin (1999) has reported. Burning these and replacing them with grassland greatly reduces transpiration and cooling. Cleared areas of the Amazon rainforest have experienced up to 30% less rainfall, and temperatures some 1°C higher, than in the surrounding forest. Similar changes have occurred in tropical and sub-Saharan Africa.

Comparable heating and drying occurs in summer-autumn with the clearing of trees and their replacement by annual crops in mediterranean climates. The deep-rooted natural vegetation of these climates transpires right through the dry summer, whereas the dry residues from winter-spring crops do not transpire at all. As well as raising summer temperatures, both locally and in adjacent wooded and coastal areas subject to hot land winds, this may be expected to delay and reduce rainfall around the autumn break of the normal rainy season. Less 'roughness' of the land surface also reduces turbulence and uplift of potentially rain-bearing winds. Pitman et al. (2004) and Timbal and Arblaster (2006) attributed to the latter factor about half the major winter rainfall decline that occurred in south-western Australia following extensive land clearing that peaked in the 1960s. The other half they attributed to a poleward shift in the weather systems that accompanied worldwide warming from the mid 1970s to the 1990s.

Contrariwise, irrigation in arid regions brings about marked cooling. Couzin (1999) cited research at Colorado State University, Fort Collins, showing that extension of irrigated crops in the 1990s locally reduced July mean temperatures by as much as 2°C. Combined with more cloud and rainfall, this allowed conifers to establish naturally on the Rockies' eastern slopes at significantly lower altitudes than before.

Finally we may examine the work of Balling and colleagues (Balling 1988, 1991; Balling et al. 1998), who studied the effects of land degradation and desertification on temperature trends. One detailed set of studies compared temperatures on either side of the Arizona/Mexico border. Strictly controlled grazing on the Arizona side maintains good vegetation cover, whereas that on the Mexican side has been chronically over-grazed and undergone marked degradation. After allowing for differences in latitude and altitude, both maximum and minimum temperatures measured higher on the Mexican than on the Arizona side, especially in summer; the warming trend over time was also greater. Comparing other Northern Hemisphere regions undergoing desertification with adjacent better-watered regions gave similar results (Balling 1991). Balling points out that some 10% of the earth's land area suffers from severe desertification, with a further 20% slightly to moderately affected.

Whether earlier land use changes have influenced global temperature remains uncertain. Govindasamy et al. (2001) suggested that the downward trend of estimated Northern Hemisphere temperatures from AD 1000 to 1900 could have resulted from replacement of woodland by grass and cropland as populations grew. But the big picture does not fit the detail, at least for Europe. The steep temperature fall of the Late Middle Ages accompanied declining population that resulted from famine, plague and war. Some arable land probably reverted to woodland. Then the renewed expansion of pastures, cropping and population from the 17th century onwards coincided with mostly recovering temperatures: again contrary to what Govindasamy et al. postulate.

It seems likely that temperature changes in both directions from evolving land use over recent centuries have largely cancelled each other out. But since the mid 20th century a new scenario has started to develop. Clearing of temperate woodlands for agriculture has reached its practical limit, as has the exploitation of water resources for irrigation. Any increase in irrigation area comes mostly through using more water-efficient methods such as drip irrigation and, now increasingly, regulated deficit or partial rootzone drying techniques. The overall cooling effect of irrigation has thus also reached its limit. Indeed, cases have emerged where over-diversion of water for irrigation has led to desertification and warming down-stream, as in the case of the Aral Sea in the former USSR. At the same time forest clearing has rapidly increased in the tropics and sub-tropics, where clearing regularly results in warming.

All climate effects from land-use changes are now in the direction of warming. Their global impact cannot be estimated accurately, but could be very significant across low and middle latitudes.

I have omitted so far any mention of local warming due to urbanization. Given the still very small proportion of the land plus ocean surface so affected, this can hardly have raised global temperatures appreciably. But effects on the *thermometer record of land temperatures* are a different matter, since most land recording stations are in or close to towns or cities; and these, in addition to their own effects, can often be in topographic situations that themselves bias temperatures. We will look at these aspects when discussing temperature measurement in Sections 12.6 and 12.7.

12.4 Modelled temperature feedbacks

Positive feedbacks amplify, and negative feedbacks reduce, the temperature changes due to primary agents such as solar output, greenhouse gases or aerosols.

Subsection 12.3.1 has already noted feedback via water vapour. Warming from any cause automatically increases evaporation and atmospheric water vapour content, which in turn serves to amplify the original warming. Likewise cooling reduces water vapour content, which amplifies greenhouse-mediated cooling. This is a classic positive feedback system, that responds purely to temperature changes provided that water is available for evaporation.

Here we are concerned more with other feedbacks that contribute much to the inherent uncertainty of climate models. Two main types that have exercised modellers are those associated with ice and snow cover at high latitudes and/or altitudes, and, more crucially, those resulting from changes in cloud cover or type. A further important avenue of feedback, to which the models have hitherto paid little attention, is direct surface cooling by water evaporation and plant transpiration. We will return to it later in subsection 12.8.3.

12.4.1 Ice and snow cover

The feedback role of ice and snow is readily understood in principle. Each reflects away much more radiant energy than water or uncovered land. With cooling (seasonal or longer-term), greater reflection from more snow or ice causes more energy loss to space by reflection and hence more cooling. Thaw following warming reverses the process and accelerates the warming. Again we have a straightforward positive feedback system.

In practice the situation is less simple. Pristine ice and snow are highly reflective, but surface accumulation of soot and other pollutants can greatly reduce this. Secondly, the main effects are localized to where ice and snow cover changes regularly with the seasons or with the longer-term climate fluctuations. Thus the climates of arctic and near-arctic northern Europe, Siberia and North America are not only highly continental, i.e. they have a wide summer/winter temperature range; they also have magnified longer-term temperature changes compared with other regions. The extreme fluctuations of the ice ages and inter-glacials probably owe much to such feedbacks as well as to astronomical factors.

How much the ice and snow feedbacks add to minor and shorter (e.g. decadal to centennial) climate changes, such as we are concerned with here, is harder to discern. They are undoubtedly significant for the middle to high-latitude Northern Hemisphere, and therefore to much cool-climate viticulture. But any effects elsewhere, and on global average temperatures, will obviously be much less.

12.4.2 Clouds

Feedback by clouds remains the biggest source of uncertainty in climate change studies.

Much of the confusion arises from the multiple and conflicting effects of cloud. Reflection and scattering of incoming short-wave solar radiation reduces that reaching the earth's surface. But cloud also has a greenhouse role by absorbing and retaining outgoing long-wave radiation. Light high cloud blocks little of the incoming radiation but effectively captures outgoing radiation, the result being a net warming. Heavier mid-level and low cloud, by contrast, is extremely reflective and causes net cooling despite having a night-time greenhouse role. It is also the source of rain and hence surface evaporative cooling. Conversely the condensation of water vapour to form cloud adds heat to the atmosphere. These diverse effects vary relatively across regions and seasons, depending on the amounts and types of cloud and the locations of cloud formation and rain.

The net current contributions of clouds to global average temperatures was totally unknown until results became available from the satellite-based Earth Radiation Budget Experiment, commenced in 1984 (Ramanathan et al. 1989). This showed, for the April 1985 period, a global average excess of cooling over warming by clouds of 13.4 Wm^{-2}, or three to four times the 4 Wm^{-2} of warming then postulated for an effective doubling of atmospheric CO_2. Over mid to high latitude oceans the excess of cooling reached as high as 100 Wm^{-2}, while by implication the lack of low cloud over dry continental interiors and some low-latitude oceans must lead to a balancing excess of heating there. The figures suggest, but do not prove, that the overall balance of feedback from clouds is negative.

A further source of cloud-mediated feedbacks may be important. Several studies (Charlock 1982; Somerville and Remer 1984; Mitchell 1993; Lee et al. 1997) conclude that in warmer and overall moister atmospheres the middle and low-level clouds will contain more liquid water, making them optically thicker and more reflective of short-wave radiation. This in itself constitutes a negative feedback.

The treatment of clouds in climate models (Cess et al. 1989, 1996) has followed a mostly opposing argument. The dominant assumption here has been that warming from any source reduces low to mid level cloud cover, leading to further warming (a positive feedback). Early models gave feedback ranging from about neutral to very strongly positive, resulting in a *three-fold* range in estimated climate responses to initial forcings. This accounted for nearly all the total variation among models. The later models (Cess et al. 1996) moderated this range somewhat, but still showed mainly neutral to positive feedbacks. No model to that date showed a strongly negative cloud feedback.

The model assumptions on cloud feedback are thus clearly open to question. In any case cloud feedback over land can hardly be considered realistically in isolation from the related hydrological factors of rain and cooling by evaporation at the soil surface or by plant transpiration, which we have already touched on and will examine further in subsection 12.8.3. The conclusion there, as seems intuitively right and consistent with common observation, is that the hydrological cycle as a whole, including clouds, serves overall to moderate externally forced temperature changes rather than to enhance them.

12.5   Attribution of causes

Subsection 12.2.4 looked at studies estimating historical (pre-industrial, pre-thermometer) temperatures from various proxy measures, based on the latter's subsequent correlations with contemporaneous thermometer records.

This section focuses solely on the period of the thermometer record from 1850 on, which happens also to be that of rising atmospheric greenhouse gases. The question to be answered is: how much of the modern temperature record can be attributed to continuing natural climate variation, and how much to anthropogenic greenhouse gases and perhaps other anthropogenic effects?

The fact that greenhouse gases over time can be fairly well estimated facilitates this process to a point, as does the conclusion that natural climate variation over longer than decadal timescales depends more or less exclusively on astronomical and solar factors. Careful statistical matching of all these against measured temperatures should theoretically show how much of temperature change has been due to each.

One approach, associated mainly with the core IPCC school of thought, has been to confine estimates of change in solar influence to those based on irradiance measures directly by satellites since 1980, and their effects on earth temperature to those estimated via current energy-balance climate models.

12.5.1 *The IPCC approach*

This takes as its starting point the crucial paper of Foukal and Lean (1990), which used satellite measurements of solar irradiance from 1980 to 1988 to construct an empirical model of irradiance variation between 1874 and 1988. They found that although irradiance, as measured by its secular (long-term) trend across the 11-year sunspot cycles, had increased steadily since 1945, the effect when fed into a mainstream climate model was a rise of only 0.02°C for the whole period. They concluded that: "solar irradiance variations are unlikely to have had significant influence on climate over the past century *unless the sensitivity of climate to solar irradiance is seriously underestimated by currently accepted energy balance models*" (emphasis added).

Many other papers then and after have delivered the same conclusion, but with less regard to the proviso. Representative examples are those of Kelly and Wigley (1990), Wigley and Raper (1990) and Hoffert (1991). Shine et al. (1990), under IPCC imprimatur, present the case in some detail (pp. 62–3), arguing that the proxy-based estimates of pre-industrial solar variations and temperatures are unreliable and to be disregarded. The IPCC (2007) continues to discount any significant solar influence on global climate. Consistent with this, the models used throughout these studies self-evidently incorporate assumed responses to greenhouse gases that leave little remaining variance for attribution to other factors. Naturally they conclude that greenhouse gases are the chief cause of recent temperature change.

12.5.2 *Wider approaches*

A second group of analyses continues to work with climate models but uses directly the records of solar irradiance as estimated from proxy data, notably those of Friis-Christensen and Lassen (1991), Kelly and Wigley (1992), Hoyt and Schatten (1993) and especially Lean et al. (1995). They increasingly include other forcing agents besides greenhouse gases and solar irradiance, including the cooling effects of volcanic aerosols and (as a new factor) anthropogenic aerosols. Greenhouse gases must now compete for influence with a growing array of alternative contributing factors. Representative papers are those of Kelly and Wigley (1992), Schlesinger and Ramankutty (1992), Tett et al. (1999, 2002), Stott et al. (2000) and Meehl et al. (2004). Most reach a common conclusion: that early 20th century temperature rise had a moderate to dominant solar input, whereas that after 1970 has been due principally to anthropogenic greenhouse gases.

Two points arising from this group of studies deserve comment. First,

Kelly and Wigley (1992) include graphs and statistical analyses of temperature responses: to greenhouse gases alone, to solar factors alone, and to the two combined. Best fit, both graphically and in total explained variance, comes from solar alone and (very similarly) solar plus greenhouse. Greenhouse gases alone gives a poorer fit. Yet the authors inexplicably conclude that greenhouse gases explain most of the temperature response.

Second, the big role of anthropogenic aerosols in the later models of this group calls for scrutiny. Anderson et al. (2003) discuss the two ways by which anthropogenic aerosol contributions have been estimated. The first is by 'forward' calculations based on aerosol physics and chemistry. Although the basic principles are fairly well understood, many uncertainties exist that largely rule out practical utility.

The second approach involves 'inverse calculations', which "infer aerosol forcing from the total forcing required to match climate model simulations with observed temperature changes". Later: "Unfortunately, virtually all climate model studies that have included anthropogenic aerosol forcing as a driver of climate change ... have used only aerosol forcings that are consistent with the inverse approach." In other words, the values for anthropogenic aerosol forcing used in the models are based, not on good evidence, but on what is needed to conform to pre-judged levels of forcing by greenhouse gases. The greater the offsetting aerosol influence, the higher the greenhouse gas forcing that can be justified.

This is an obvious and scientifically unacceptable piece of circular reasoning. Models so constructed cannot be relied on to project future climates.

A third point arises from several later papers. Stott et al. (2003) examined treatment of the variance attributed to different forcing factors in the existing models. They found that the regression techniques used, in much the same way as in back-projections of pre-industrial temperatures (see subsection 12.2.4), depressed the inferred solar contribution by a factor of two or more, at the same time amplifying that of greenhouse gases. Scafetta and West (2005 and references therein) revealed comparable model biases against solar-related contributions.

The emerging evidence of deficiencies in established climate models led to a third, and to date final, group of analytical papers which largely dispense with theoretical models and look instead at the pragmatic correspondence between solar and cosmic ray phenomena, anthropogenic factors, and the measured temperature record. Notable papers in this group are those of Lean et al. (1995), Lean and Rind (1998), Solanki and Krivova (2003) and Scafetta and West (2005, 2006). They reach remarkably unanimous conclusions. About half of the century's recorded temperature rise of about 0.6°C can be attributed to the sun's fluctuating but overall rising irradiance and magnetic field. Close to three-quarters of that pre-1950 can be so attributed, but at most only a quarter to a third of that post-1970. The remaining half must then stem from anthropogenic factors, which are relatively weak early in the century but accelerate rapidly towards its end. The combined fit to the temperature record is good. Later sections will examine the nature of the anthropogenic contributions.

12.5.3 The question of solar enhancement

Part of modellers' scepticism over a contribution of solar changes to climate fluctuations has come from a lack of known possible mechanisms. Disregarding the historical and proxy evidence, there is no proven theory as to how very small changes in solar output can cause the temperature changes observed at the earth's surface. That raises the question of whether mechanisms may exist that amplify the direct effects of solar variations on climate. Two such mechanisms, not mutually exclusive, have been suggested.

Haigh (1994, 1996, 2003) pointed out, like Fairbridge (1962), that while changes in total irradiance are small, those in the ultra-violet part of the spectrum are relatively several times greater. Ultra-violet energy is absorbed in the mainly upper atmosphere, where it causes oxygen dissociation and ozone (O_3) formation, which in turn increases UV absorption. Resultant warming in the middle to upper atmosphere changes temperature relations between altitudes. Haigh hypothesizes that this influences the major global wind systems, causing a small but significant poleward displacement of weather systems and a magnification of warming when solar irradiance is highest. Shindell et al. (1999) produced model evidence supporting this thesis, albeit with only a small estimated effect on global temperature overall.

The second suggested mechanism, which remains controversial, is that of the Danish space researcher Henrik Svensmark and his colleagues (Svensmark and Friis-Christensen 1997; Svensmark 1998). This suggestion grows from the well-established inverse relationship between the intensity of cosmic radiation reaching the earth and sunspot incidence, the latter correlating with the intensity of the sun's magnetic field that shields the earth from cosmic radiation (see subsection 12.2.3). The theory is that ionizations resulting from bombardment by cosmic particles contribute to the formation of cloud condensation nuclei, an idea later supported by direct experimental evidence (Svensmark et al. 2007). Reduced sunspot incidence and solar magnetic field strength, then, are supposedly associated with greater cosmic particle penetration, more cloud condensation nuclei, and more, and more reflective, clouds. Other studies (Marsh and Svensmark 2000; Yu 2002) suggested that the main increase is in low cloud, causing maximum cooling. Likewise Usoskin et al. (2004) found that inter-annual variability of low cloud increases poleward in a consistent relationship with measured ionizations, again suggesting that the latter could modulate cloud.

Many immediately questioned Svensmarks's thesis, e.g. Kernthaler et al. (1999), Kristjánsson and Kristiansen (2000), Jorgensen and Hansen (2000), Wagner et al. (2001) and Kristjánsson et al. (2002). Later literature, however, has generally received it as credible if unproven.

Neither mechanism has been recognized in mainstream climate modelling or by the IPCC.

12.5.4 Conclusions and comments

We now have what seems a realistic attribution of causes for the 20th century rise in globally recorded temperatures. Approximately 0.3°C, or half the total, results from changes in solar irradiance and magnetic field, consistent with the longer historical record as inferred from a variety of sources. Volcanic aerosols play a part within the period, a substantial absence of them probably contributing to the rapid warming of the 1920s to 1940s and a succession of eruptions helping to explain the slight cooling from the 1950s to early 1970s; but these cancel each other out over the whole period.

The remaining 0.3°C can be taken as representing the net total of anthropogenic effects. These include greenhouse gases and anthropogenic ozone pollution (both warming), and anthropogenic aerosols (mostly cooling?) which to a probably very small extent may counteract the first two. But we must also include among anthropogenic effects those of land uses as described in subsection 12.3.3; those of historical trends in thermometer location and immediate surrounds, to be discussed in subsection 12.6.3; and finally, those of the warming peculiar to the mainly city or town environments of the recording stations. That will be the subject of Section 12.7.

Finally, it is appropriate to point out two extremely questionable procedures used by the IPCC (2007) in constructing its case for negligible solar influence, and from that a predominant role of anthropogenic greenhouse gases.

First, it arbitrarily selects 1750 as the base year for comparing solar irradiance with that of 2005. But the best evidence is that the mid 18th century was itself at a temporary peak of irradiance, comparable to that of the late 20th century (Friis-Christensen and Lassen 1991; Lassen and Friis-Christensen

1995; Hoyt and Schatten 1993: see Figure 12.4). The same studies show the late 19th–early 20th century, which is the starting point for most models and attributions, as being at an irradiance low point, as was the closely related strength of the sun's magnetic field (Lockwood et al. 1999). All estimates show the sun's irradiance in the early 19th century, at the true start of industrial expansion, to have been even lower, having fallen steeply from its mid 18th century peak. These, not 1750, are the valid base points for modelling recent solar effects on temperature. They would show a much larger contribution. The IPCC totally ignores the extensive, and largely unanimous, research described in subsection 12.5.2 that ascribes about half of the 20th century's thermometer-indicated temperature rise to solar factors.

Second, the IPCC assumes a global cooling by anthropogenic aerosols that offsets almost half its estimate of warming from anthropogenic greenhouse gases. Yet as we saw in subsection 12.3.2, although there is evidence for seasonal local cooling in and downwind of the areas of main sulphur emissions, that for a significant worldwide effect is at best weak.

Each of the two procedures approximately doubles the warming attributable to greenhouse gases. Without them, the estimate of past, present and future anthropogenic warming would be up to four times less than the IPCC claims, bringing it into agreement with the pragmatic estimates of the previous subsection.

How can such a conclusion be reconciled with the basic physics of greenhouse gases as incorporated in the climate models? Assuming the physics to be correct, the most likely explanation is through negative natural feedbacks such as through the hydrological cycle, that the models do not adequately take into account. Indeed, most models seemingly still assume net positive feedbacks from the hydrological cycle. Another avenue, for probably slow-acting negative feedbacks, lies in ocean dynamics and their coupled effects on the wind systems, such as discussed by Cane et al. (1997). These are still incompletely understood, but have undoubted capacity to influence climate and act as feedback agencies.

The second main possibility is that direct responses to greenhouse gases, especially CO_2, have been wrongly calculated. I am not competent to discuss the physics of this relationship. However it may be observed that where, as in this case, a response is so near to saturation, the result can be excessively sensitive to experimental errors or differing assumptions.

We come back, then, to the best available analysis of temperature over the last century or century and a half as actually recorded. Of about 0.6°C recorded temperature rise, half can be ascribed to variations in solar irradiance and magnetic field, and the rest probably to anthropogenic causes. But much of the latter can be shown to derive from mistakes and biases in the thermometer record, to which we now turn.

12.6   Thermometer records 1850–

12.6.1 *The 'consensus' record*

Three research groups have monitored in detail the world trends of surface-recorded temperatures since the mid 19th century: the Climatic Research Unit at the University of East Anglia, Norwich, England; the NASA Goddard Institute for Space Studies, New York; and the Hydrometeorological Institute of the Former Soviet Union (P.D. Jones 1994). The results show mostly good agreement. Here we focus on the University of East Anglia record, which is generally taken as authoritative and has been the most used for climate modelling and attribution. I shall term it the 'consensus' record.

First detailed accounts from this group are those of P.D. Jones et al. (1986a) for the Northern Hemisphere, and P.D. Jones et al. (1986b) for the Southern Hemisphere, both based on land temperatures only. P.D. Jones et al. (1986c) incorporated marine records and summarized the results on a global basis. In doing this, however, they rejected the sea surface measurements and resulting adjusted values of Folland et al. (1984) (to be described in the following subsection). Rather than adjusting directly for evaporative cooling of the water samples, as Folland et al. had done, they derived time-related adjustment factors by comparing the raw water sample readings with those from adjacent land.

Resulting adjustments to the raw sea surface values were as follows: 1861–1889, +0.08°C; 1903–1941, +0.49°C; 1942–1945, –0.10°C; 1946–1979, nil; with linear interpolation between 1889 and 1903.

The important thing to remember is that this procedure changed the time series of the marine data to conform to that of the land data. The resulting shape of the global time/temperature graph therefore reflects that of the land, not of the oceans or a true combination of the two. The distinction is crucial for estimating global climate change, as later discussion will show.

P.D. Jones et al. (1999), P.D. Jones and Moberg (2003) and Brohan et al. (2006) give global and hemispheric updates. They use differently based correction factors for the raw water sample readings, based on the later analysis of Folland and Parker (1995) (see following subsection). The new factors largely repeat those of Jones et al. cited above. Figure 12.6 shows the Jones and Moberg graphs, which cover the years 1860–2000.

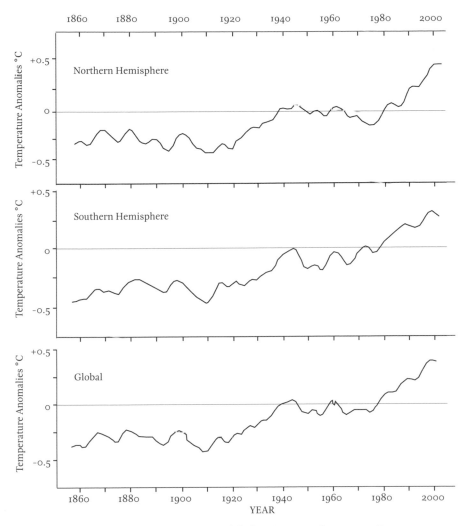

Figure 12.6. Revised hemispheric and global estimates of mean surface temperatures 1858–2001, expressed as anomalies from the 1961–90 average means. From P.D. Jones and Moberg (2003).

These graphs, as do the others of the 'consensus' series, show relatively steady temperatures between 1860 and 1880, small dips around 1890 and 1910, a rise of some 0.4°C between 1910 and 1940, an uneven fall of perhaps 0.1°C from then to about 1975, then a steep rise to 2000, averaging nearly 0.5°C globally but a little more marked in the Northern than in the Southern Hemisphere. The authors acknowledge uncertainties, chiefly from paucity and unevenness of coverage in the 19th century, and in some remote regions up to the present. Also the earlier records show vexing signs of disparity between the land and marine measurements, as described above. These must be explained and reconciled before global climate change can be properly measured or understood. That in turn demands an analysis of both sets of records, to see which (if not both) might be in error.

12.6.2 Land versus ocean records

Land records come from fixed-location thermometers, mostly in standard-height screen shelters such as the Stevenson screen. However changes in screens, thermometers and location over the years can create discontinuities, while local factors of topography, surrounding vegetation, ground surface and land use, including urban development, can substantially influence readings and their course over time.

P.D. Jones et al. (1986a) describe detailed measures taken to detect discontinuities caused mainly by station shifts, and measures to allow for them. They also claim in the process to have successfully identified stations showing urban warming, and to have removed them when estimating land temperature trends. But they removed only 38 out of 1,584 Northern hemisphere records on this ground, of which 31 were from North America. This seems unbelievable, given that many of the non-American locations were large towns and cities where the effects of urban warming were already well known. Moreover, a critical review by Wood (1988) showed that the 22 locations excluded in the USA covered population sizes from 108 to 1,600,000, with no apparent relation between population and exclusion. Whatever the basis for exclusion, in few cases could it have been urban warming as commonly understood. The land temperature updates of P.D. Jones et al. (1999) and P.D. Jones and Moberg (2003) reportedly continued to rely on the same screening method (Gallo et al. 2002).

But these are matters relevant mainly to the later temperature record, as will be discussed fully in Section 12.7. Urban warming can have played little part in the mid 19th century, which is the period we must examine first.

The first comprehensive analysis of worldwide marine temperatures, of Folland et al. (1984), comprises two parallel sets of measurements: those of air temperatures at ship deck height, and those of near-surface water as collected in immersed buckets; or later, from about 1941, mainly via engine water intakes.

It was early recognized that daytime deck measurements of air temperature, even in properly shaded screens, were unreliable because variably influenced by solar heating of the ship's structures. This led most researchers to the exclusive use of night temperatures for establishing trends over time.

Even then, increasing deck heights above sea level introduced a time-related bias. Folland et al. estimated and used corrections for this from established atmospheric physics, ranging progressively from –0.13°C for measurements before 1890 to +0.09°C for 1976–81.

Theoretically the temperature of near-surface water (sea surface temperature, or SST) should be a more reliable measure over time. It is essentially uniform through the top few metres of well mixed water, while water's high specific heat gives strong short-term thermal inertia. The problem is in the method of sampling. Prior to engine intake sampling, the buckets used were cast on ropes over the ship's side, then hauled back to deck and the water's temperature taken. Throughout the process evaporation took place which resulted in cooling. This would vary depending on air conditions, time taken in bucket raising and temperature measurement, and the insulation and evaporative properties of the bucket. Despite these uncertainties, by matching the full record of sea surface temperatures against that of adjusted night air temperatures, Folland et al. found that the bucket sample cooling could be adequately corrected by adding 0.3°C to all sea surface measurements up to 1940, i.e. up to the universal adoption of engine intake sampling.

So adjusted, estimated sea surface temperatures from the mid 19th century to about 1880 closely resembled those of the 1951–1960 control period. There followed a gradual fall of 0.2°C to the end of the 19th century, then a further steep fall of 0.3°C over the next few years. Temperatures rose again between 1910 and 1940 by about 0.5°C to regain their mid 19th century level, after which they remained more or less stable to 1980 apart from a peak in the early 1940s that was probably related to a strong El Niño at the time.

This is a pattern clearly different from the land-based measurements of P.D. Jones et al. (1986a, 1986b) or the supposed hemispheric and global averages of P.D. Jones et al. (1986c) and P.D. Jones and Moberg (2003), shown in Figure 12.6. Importantly, it shows mid to late 19th century marine temperatures some 0.3°C higher than expected from the corresponding land measurements.

Attempts since to refine the interpretation of early sea surface temperature measurements have hinged on the types of bucket used, which differed greatly in their thermal properties. Folland and Parker (1995) give a detailed account. It is known that nearly all in use after 1900 were of canvas, and therefore poorly insulated and highly subject to evaporative cooling; but opinions have varied as to the types used earlier. Official policy in the mid 19th century appears to have favoured wooden buckets, which being less evaporative and much better insulated would have cooled less. However they were reportedly unpopular, partly because of their weight and partly because their metal staves scratched the ships' paintwork. No known evidence has survived as to how much they, or alternative bucket types, were used. Attempts to factor them in when estimating 19th century marine temperatures are therefore speculative.

In their own estimates, Folland and Parker (1995) assigned corrections to the raw bucket data rising from only +0.11°C in 1856 to around +0.40°C from 1920 to 1941, falling to nil after 1941. Subsequent updates of the 'consensus' temperature record (P.D. Jones et al. 1999; P.D. Jones and Moberg 2003; Brohan et al. 2006) used these corrections. The resulting pattern resembles that of P.D. Jones et al. (1986c) based on compliance with the land records, and necessitates an implied exclusive early use of wooden buckets. But with no firm evidence on that point, their conclusions remain questionable.

The differing land and marine time/temperature estimates lead to contrasting conclusions on the causes of climate change. On the one hand those of Folland et al. (1984) suggest a fluctuating climate with little long-term trend. Admittedly it does not take in the 20th century's last two decades when undoubted warming took place. But for the period covered, the fluctuations correlated best with proxies for purely natural agents of climate change and do not show overall warming.

By comparison the 'corrections' of P.D. Jones et al. (1999) and their successor series depress the mid 19th century marine temperatures but raise those of the early 20th century, resulting in a clear overall rising trend, albeit with some fluctuations, that gives a fair fit to rising atmospheric greenhouse gases. It is these estimates that have since provided the temperature basis for nearly all reported modelling and attribution of climate change causes.

But later studies of marine temperatures by Kaplan et al. (1998), Smith and Reynolds (2003) and Rayner et al. (2006), with the advantage of more extensive data and analysis, have largely confirmed the original conclusions of Folland et al. (1984). Figure 12.7 shows the Smith and Reynolds graph. All their sea surface temperatures show moderate late-mid 19th century warmth, at most only about 0.1°C below that of the mid 20th century and some 0.2°C above equivalent recorded land temperatures of the time. The marine temperature dip between 1880 and 1910 is also largely confirmed, now about 0.4°C compared with 0.5°C in Folland et al. Thereafter the respective land and sea records generally agree up to the end of the Folland et al. record.

However, the later data of Kaplan et al. (1998, Figure 14) show a further divergence, with the land estimates warming relative to those of the ocean by

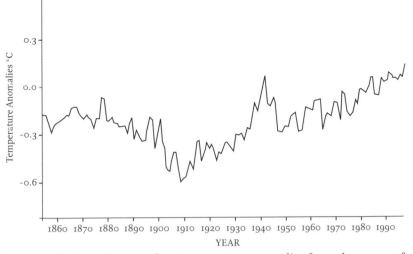

Figure 12.7. Estimated sea surface temperature anomalies from the average for 1954–1997, between latitudes 60°S and 60°N. From Smith and Reynolds (2003).

about 0.1°C over the century's last two decades. Taking all figures together, we can see that the newly estimated marine record warms only 0.3°C from 1850 to the end of the 20th century, compared with 0.7°C for the land record.

Such a developing discrepancy between land and sea trends over a long period should not exist, given the reasonable expectation of a dynamic equilibrium between adjacent land and sea temperatures. It is a clear indication of developing bias over time in one or other (or both) of the temperature series.

The following discussion will show beyond reasonable doubt that the fault lies primarily in the land record. Part of this undoubtedly comes from inadequate correction for growth in urban warming (Section 12.7). But our study of topographic adjustments for defining viticultural climates (subsection 2.1.5 and Chapter 11) suggests that another historical trend has also contributed.

12.6.3 *Effects of land topography*

Early land temperature records came, more or less by definition, from old-established settlements. For historical and practical reasons most were from river valleys, where fresh water was available, river transport possible, and the alluvial soils conducive to agriculture and horticulture. Often towns and cities were at river crossings or at upper limits for navigation, which were natural sites for commerce to develop. Ocean ports likewise tended to be at or close to the mouths of rivers, or at least at the foot of valleys affording natural harbourage.

Table 2.2 shows that such sites are typically subject to cold night air flow and sometimes ponding, which depresses their minimum temperatures and means compared with the landscape overall. Geiger (1966) discusses the relevant mesoclimatology in detail. Assuming urban warming still to have been negligible in the mid 19th century, this provides a simple explanation why the early recorded land temperatures show a cold bias relative to those of the adjacent sea.

For near-coastal land sites such as presumably featured in the land/sea matchings of P.D. Jones et al. (1986c), a further factor would have operated. The valley-floor sites still receive the coldest air of the morning land breezes, but also they get the earliest and strongest afternoon sea breezes funnelling up the valleys. On both counts their mean temperatures are depressed relative to the average for the landscape. And it is the *average* of the landscape temperatures, adjusted for altitude, that should be in equilibrium with adjacent sea surface temperatures.

Later land settlements developed with fewer constraints. Rail and motor transport, piped or newly conserved water supplies, and artificial fertilizers that facilitated agriculture on poorer upland soils – all combined to allow new or faster growth of settlements away from the valley floors.

Some were satellite settlements, falling within the thermal zones of a valley's middle and lower slopes; they would on average have enjoyed markedly better night air drainage and warmer nights than their parent settlements on the adjacent valley floors.

In due course many of these newer settlements acquired weather recording stations of their own. The predictable result through the early and middle years of records is an increasing mesoclimatic weighting towards warmer sites. This leads to a spurious overall warming trend with time in the altitude-adjusted thermometer record, that is independent of but adds to that of urban warming.

12.7   Urban warming and summary of biases

The reasons for specifically urban warming are straightforward enough. Two main mechanisms are at work.

First, paving and brick or stone building materials absorb heat from sunlight, store it and re-radiate it at night. Daytime air temperatures are little affected, but re-radiation, most marked in the evening but continuing through the night, raises night and minimum temperatures and therefore mean and average temperatures. Diurnal range falls.

Second, industrial and domestic use of power or fuels brings about local warming directly, in proportion to their extent and intensity of use. We know little about the size of this effect, but can be confident of its direction.

One can reasonably assume that the direct contribution of urban warming to global temperatures is negligible, although it may be regionally significant in some densely populated areas. The combined total area of open country-side and the oceans overwhelmingly dwarfs that of the towns and cities.

What we are concerned with here is different. It is the contribution of urban warming to the *thermometer record* of land temperatures, which has been the main basis for estimating land temperatures overall and which, as we have seen, has dominated the estimation of global temperatures over the last 150 years. The fact that a high proportion of land thermometer records come from variably urbanized sites means that a spurious warm bias will contaminate the record unless properly estimated and discounted. Moreover, the bias will grow as urban centres grow and their energy use intensifies.

Kukla et al. (1986) reviewed literature-reported differences in rates of warming between major centres (populations mostly well over 100,000) and neighbouring small centres with populations typically under 7,000. Individual comparisons differed widely, but overall the large centres warmed 0.12°C per decade faster than the small ones. The authors noted that this probably under-estimated total urban warming, since even small centres would have had some warming of their own relative to the open countryside.

Karl et al. (1988) sought further to quantify urban warming differences among small centres and against rural sites within the USA, the defini-tion of 'rural' here being population less than 2,000. While acknowledging wide differences in thermometer environments unrelated to population, they developed estimates of average warming bias based on population, as being the only then universally available measure of village, town or city size. Table 11.2 in the previous chapter sets out their estimates across the full range from the smallest rural centres to the largest cities. But where comparisons were confined to neighbouring pairs of rural and mostly small urban sites that comprise the US Historical Climatology Network, they concluded that urban warming could account for only 0.06°C of the network's total warming warming between 1901 and 1984, or perhaps up to 0.09°C relative to the open countryside. "However, the use of another dataset with greater urban representation, if uncorrected, could easily result in much larger biases."

Estimates of urban warming for the USA do not necessarily apply else-where. Torok et al. (2001) compared them with data from Europe and Australia. The urban warming bias rose somewhat less steeply with popula-tion in Europe than in the USA, and a little less steeply again in Australia.

The authors attributed this to denser populations and especially more vertical city architecture in the USA, as compared with dispersed and lower-profile Australian cities, with Europe intermediate. Differences in intensity of energy input might also have contributed. But whatever the geographic differences, it is clear from the evidence that most settlements have some degree of urban warming.

Hansen and Lebedeff (1987) and Karl and P.D. Jones (1989, 1990) further examined data for the world and the USA respectively. Based on 20th century temperatures up to 1984, they estimated that urban bias could account for between 0.1 and 0.2°C of the recorded rises for the century. The detailed later study of Hansen et al. (2001) estimated a figure for the USA of 0.15°C.

A number of researchers have criticized these figures as excessive. P.D. Jones et al. (1990), in a study since widely cited as authoritative, followed the relative warming rates of urban and 'rural' sites in the former USSR, China and eastern Australia and found no difference. They suggested that the difference of 0.15°C previously reported for the USA was atypical, and concluded that the urban warming component for the global land record could not exceed 0.05°C per century, i.e. an order of magnitude less than the total recorded warming.

Later research by T.C. Peterson and colleagues went further. Peterson et al. (1999) followed the trends of mean temperature among the 7,280-station time series of the Global Historical Climatology Network (GHCN), the stations being classified urban or rural from a combination of maps and satellite-observed night-time lights. They found little, if any, difference in either temperature or its time trend between urban and rural sites, and concluded that: "the ... global temperatures time series from in situ situations is not significantly impacted by urban warming."

Peterson (2003) then compared 40 clusters of 'urban' and surrounding 'rural' sites across the USA, the definition of rural being that they were associated with centres of less than 10,000 population. Data treatment included established standard adjustments for site latitude and altitude, times of re-setting maximum/minimum thermometers, and additionally type of instrument and its housing, which tended to differ between urban and rural sites and thus introduced a systematic bias. The unadjusted data showed urban sites to average 0.31°C warmer than those classed as rural; but after all adjustments, this difference was reduced to 0.04°C. Peterson concluded that the data showed no evidence of an urban warming effect, and therefore adjustments in the time/temperature series to account for urbanization "are not appropriate".

The argued existence of at most only a very small urban warming bias in land temperature trends has become widely accepted, e.g. Brohan et al. (2006). The IPCC (2007) cites a figure of <0.006°C per decade, which it describes as negligible. Gallo et al. (2002) noted that the Global Historical Climatology Network currently made no routine adjustments for urban warming.

We are thus presented with a paradox. The more or less universal existence of urban warming is well established from both temperature measurements and physical principles. Yet official climatological opinion now discounts it, together with any significant effect on the time/temperature record despite widespread increases in urban populations and/or their intensities of energy use. How can this be resolved?

Subsequent work of Peterson and Owen (2005) provides a possible partial answer. They conducted a more detailed study having the same scope as that of Peterson (2003), using alternative sets of metadata to classify recording sites as urban or rural: firstly satellite-observed nightlights as proxies for populations, and secondly census-derived populations across 1 km × 1 km grid squares. Nightlights proved to give a poor indication of population, and no correlation with temperatures. Gridded populations did correlate with temperatures, although correspondence to the weather station sites, defined only by degrees and minutes of latitude and longitude (1.9 × 1.4 km) could not be exact. Clearest temperature differentiation came by defining urban as having 30,000 or more population within a 6 km radius. Urban sites, so defined, averaged 0.25°C warmer than the remaining 'rural' sites after full data adjustment.

Linear regression across the whole range of population densities in this study showed a temperature spread of about 1.0°C. The authors maintained that this figure does not represent the data well; but it does seemingly agree broadly with the earlier estimates of Karl et al. (1988) shown here in Table 11.2. At the same time the data showed that population density and distribution accounted for only a small part of the individual temperature differences. A clear implication is that previous studies relying on satellite-observed nightlights, or at best crude approximations of populations and their distributions, cannot adequately identify either local or (to some extent) the more general effects of urban warming. Conclusions drawn from them must therefore be treated with caution.

Resolving the paradox further calls for a detailed analysis of meso- and micro-climatic influences on thermometer recordings, which the main available historical data do not allow. Here the detailed discussion of Peterson

(2003) has great value, reinforcing as it does what the experience of grapevine terroir tells us. Peterson argues that, other than in large and highly built-up centres, the terroir factors that most influence thermometer recordings are strictly local, e.g. site air drainage, nature of the nearest ground surface, nearby presence of trees, watered gardens, ponds, lakes and rivers, and the nature, distance and direction of buildings and paving that are sources of night warmth. These can easily override the limited broader-scale urban warming of small to medium-sized centres and of thinly populated suburban areas. And to the extent that many of the operative factors vary similarly among centres, they might at first be thought to introduce no systematic bias into the combined temperature record.

But systematic biases of several kinds cannot be ruled out. In subsection 12.6.3 we noted the tendency for old-settled towns and cities to be in major valleys, and that this could explain their recorded 19th century cold bias of about 0.2°C relative to adjacent sea surface temperatures.

In time this cold bias would have been overtaken and ultimately reversed by urban warming from city growth and increasing intensity of city energy use. Karl et al. (1988) demonstrated this for the USA, finding urban centres to be on average cooler than rural sites prior to 1925 but becoming warmer than them thereafter. This, as already explained, could be largely because the smaller and newer recording centres, away from the old cities and towns, would tend more to be on sloping or hilly country, with better night air drainage and therefore relatively warmer nights.

The results of Peterson (2003) are readily explained in the latter terms, since they were based on clusters of smaller, 'rural' sites surrounding major towns and cities.

Working from general mesoclimatic principles, we can argue further that the processes of urban warming over time will themselves have a topography-related bias. Valley-floor sites, even with extensive urban warming, have periodic influxes and ponding of cold night air from outside the area of warming, as illustrated, for instance, in Comrie's (2000) climate study of Tucson, Arizona. Peterson (2003) cites further reasons why city thermometers are often so affected, many being sited in cool and low-lying parks. Being also mainly in areas of safely high rainfall or water availability, they are nearly always well watered. Such factors not only reduce the thermometer expression of existing urban warming; they also reduce the further expression of it with urban growth.

The opposite obtains for sites with sloping or, especially, convex topography and those in valley-side thermal zones. Water availability tends to be less and more parks are dry. Any true urban warming is here fully expressed and may be enhanced. Descending warmed air from the night convections of neighbouring cities may add to this is some cases.

It therefore comes as no surprise that measured warming in rural centres can be as fast as in nearby larger urban centres: the more so when it is considered that the local factors that so much influence thermometer readings can change over time as much in small as in large centres, sometimes more.

We can now see why the methods of P.D. Jones et al. (1986a, 1990) and Peterson et al. (1999), claimed to be able to detect urban bias, have failed to do so. Firstly, as described above, they do not adequately distinguish urban from rural (i.e. less urban) localities; while within both, the extreme ecological variability of immediate thermometer surrounds so dominates the temperature records that establishing other than major urban effects is very difficult.

Further, we must query the most fundamental assumption of these and other cited studies: that lack of measured difference between large and small centres in their time/temperature trends disproves the existence of urban warming, and of resulting bias in the record.

The assumption is illogical. Just as or more probably, in the light of the ecological evidence, it means simply that both are warming at similar rates. This is not inconsistent with established temperature differences associated with large differences in urbanization or local population as shown in Table 11.2. But more importantly, it means that urban warming bias in the land record is likely to be more general and overall stronger than any of the cited studies has suggested.

It is this historical and ecological source of bias towards warming in the land record that mainstream climate researchers have failed to detect. Indeed, their approach and interpretations have served to conceal it.

One final and widely cited study that purports to disprove any influence of urban growth needs now to be addressed. Parker (2004) compared 50-year worldwide trends of minimum temperatures from days and times recorded as windy against those from days and times recorded as calm. Finding no trend difference, he concluded that: "the observed overall warming is not a consequence of urban development."

But we know that wind can have contradictory effects on minimum temperatures. Certainly it can blow urban-warmed air away, although by the time of minimum temperatures much of the night-warming effect has usually dissipated. Just as important is that it prevents night ponding of cold air, and thereby raises recorded minima at many sites. Also, light winds can be needed to move air warmed by buildings, paving etc. to thermometers typically located away in parks or other open spaces. Overall these factors tend to cancel each other out.

The only directly related study I have been able to find is that of Morris et al. (2001), who found that intensity of the urban heat island did tend to diminish once wind speed exceeded 2.0 m sec^{-1}; but the effect was at most very small, only in proportion to the *fourth root* of wind speed. Such a weak relationship, if it exists at all, could hardly register in a study such as Parker's. His results and conclusion are clearly meaningless.

We can now return to the comparison of recorded land and marine temperature (subsection 12.6.2) and to Figure 14 of Kaplan et al. (1998). Compared with sea surface temperatures, measured land temperatures gain by 0.3 to 0.4°C between the late 19th century and 1991: from 0.2°C or more below in the beginning to 0.1°C above at the end. Faster warming in the land record becomes especially marked after 1975. Probable reasons can be summarized as follows.

1. The initial low land temperatures were depressed by recording stations being disproportionately on valley floors that were sites of early city growth, and were colder than the landscape generally due to night cold air collection.

2. Warming due to urban growth progressively overtook and finally exceeded this cold bias, which nevertheless still limited the extent and rate of warming as recorded at those sites.

3. Smaller and newer centres were more likely to be away from the valley floors and to have better night air drainage, allowing full expression of urban warming.

4. A growing intensity of urban energy inputs must add further warming beyond that accounted for just by population. In one study of the Vienna climate over 1951–1995, during which its population remained static, Böhm (1998) found urban warming trends of 0.2 to 0.6°C (depending on immediate thermometer environments) as compared with the surrounding countryside. He ascribed the warming to changing urban morphology and energy consumption.

5. As discussed in subsection 12.3.3, broad-scale land-use changes appear to have moved from being historically balanced in their warming and cooling to become chiefly warming, mainly through overgrazing, desertification and clearing of tropical forests. Note that while this is not strictly a source of thermometer bias like the others, it still adds to anthropogenic warming in the land record other than through greenhouse effects.

If we accept, as I believe we should, that sea surface temperatures give the best indication of global temperatures and their changes, it follows that the cumulative anthropogenic warming bias in the land record over the last 150 years has now reached at least 0.3°C. That is half the total thermometer-recorded rise.

The recent study of Kalnay and Cai (2003) produced a similar figure for land thermometer bias from evidence of a different type. These authors reconstructed US land surface temperature trends over the period 1950–2000 by re-analysing the record of parallel atmospheric temperatures. As expected these showed no influence of urban distribution below. The resulting trend and that of direct surface measurements diverged progressively, giving an estimate of warming bias in the surface record of 0.27°C per century. This was almost double the previous higher estimate for urban warming bias of 0.15°C per century, suggested by Hansen et al. (2001) and some earlier authors. But Kalnay and Cai point out that their analysis takes in not only specifically urban warming, but the full range of associated other land use changes as well, to give a comprehensive estimate of anthropogenic bias.

The widening divergence in the late 20th century between land surface and atmospheric temperatures (Kalnay and Cai 2003) and between land and sea surface temperatures (Kaplan et al. 1998) is of particular interest. The same divergence is clear from Figures 3 and 4 of the IPCC (2007), which show a strongly accelerated rise in land temperatures, but sea surface temperatures rising at only half the rate and sea level rise accelerating only a little.

Much has been made of the close parallel between the recent rapid rise in recorded land temperatures and that in atmospheric CO_2, the general assumption being of a direct causal connection. We can now see that this is far from being necessarily so. The rise in atmospheric CO_2 results from burning fossil fuel for transport and largely urban industrial and domestic power use, together with associated deforestation and soil cultivation. All these correlate closely with growth in world population and particularly its urbanization. Their combined direct effects provide as good as or better an explanation for the anthropogenic part of recorded land temperature rise than does greenhouse warming. The latter, as far as it is genuine, should in any case be fully expressed in the sea surface temperatures. The excess rise in measured land as compared with sea surface temperatures is thus a direct measure of spurious anthropogenic biases in the land record.

There can be no doubt that the last three decades of the 20th century saw a significant rise in true global temperatures, as witnessed by recorded marine temperatures, retreating glaciers and diminishing arctic sea ice. Such medium-term rises are not new. That between 1910 and 1940 was similar (see Figure 12.6), and there is evidence of comparable temporary rises in earlier centuries: parts of a natural climate fluctuation that is undoubtedly continuing.

How much warming, then, can justly be attributed to anthropogenic greenhouse gases? Taking all evidence into account, the proven amount is: none.

The measured fluctuations in sea surface temperatures accord very well with those of natural factors, mainly solar. On best circumstantial estimate these account for about 0.3°C warming over the last 100–150 years, which represents a close to final recovery from the cooling of the Little Ice Age. Combined with the spurious anthropogenic biases described above, they and land use changes can fully account for the thermometer-recorded temperature rise of the last century and a half.

But let us, for argument's sake, liberally assume that one-third, or 0.2°C of the thermometer-recorded rise, did result from anthropogenic greenhouse gases. That must represent an upper limit. It equates roughly to 0.4–0.5°C from any doubling of CO_2 or CO_2 equivalent, which is only one-fifth of the warming predicted by the more recent greenhouse models.

How does such a discrepancy come about? Of the two major possibilities, that of unrecognized negative feedbacks from initial warming seem highly likely as already discussed, particularly those from clouds and evaporative cooling at the earth's surface. The second, as discussed briefly in subsection 12.5.4, is that the assumed basic response to added CO_2 was wrong. Any combination of these possibilities could result in the discrepancies seen here.

From a viticultural viewpoint we can conclude that any anthropogenic changes to mean temperatures will be small and, for some decades to come, unlikely to have major effects beyond those of natural climate variability. On the other hand steadily increasing CO_2 may well have significant other effects, both through diurnal temperature range and directly on the vines and fruit, some of which could be beneficial. Section 12.8 to follow will examine diurnal range and some related aspects including cloudiness and surface evaporative cooling, and Section 12.9 the direct effects of rising CO_2 on plants generally.

12.8   Diurnal temperature range, cloudiness and evaporation

12.8.1 *Diurnal range*
Nearly all temperature data generally available up to the 1980s were as

monthly or annual averages of the daily means. Only then did diurnal range and its changes start to attract serious climatological study, beginning with that of Karl et al. (1984, 1986) in the USA. Their examination of records from 130 land stations across the USA and Canada over 1941–1980 showed a progressive narrowing of the range. Mostly this was due to rising minima, with maxima unchanged or slightly falling. Excluding 24 fast-growing towns with suspected urban warming made little difference. Later studies extending to 1990 in Australia (Balling et al. 1992) and New Zealand (Zheng et al. 1997), together with global surveys by Karl et al. (1991, 1993) and Easterling et al. (1997) suggested the trend to be world-wide and that it continued to about 1990.

Appendix 1 details a study I have made of minimum and maximum temperatures across French viticultural regions as averaged for the periods 1931–60, 1961–90 and 1973–93. All records were from regional airports, which should ensure comparable topography and, in most cases, reasonable freedom from urban warming bias.

Between the first and last periods, measured annual average mean temperature across all sites rose by 0.15°C, while that for the April–October vine growing season stayed unchanged. In contrast there were big reductions in the diurnal range, by 1.1°C for the whole year and 1.2°C for the growing season; over the latter, average minima rose by 0.6°C and average maxima fell by the same amount. Comparison of the 1961–90 and 1973–93 figures suggested that the narrowing of diurnal range accelerated during the final years. The trends were broadly consistent over the country, but differed according to the season of year.

It is worth noting that the apparently accelerating reduction in maximum temperature and diurnal range in these data could hardly have resulted from increasing anthropogenic aerosols. By this time sulphur emissions in Europe were already falling (Engardt and Rodhe 1993), and this will have been especially so in France with its progressive conversion to nuclear power generation.

Later worldwide data of Vose et al. (2005), covering the years 1950–2004, confirm an apparent end to falling diurnal temperature range around 1990, but do not suggest a reversal.

12.8.2 Cloud cover changes

Doubt exists as to the accuracy of cloud cover records prior to the 1950s, due to confusions over recording methods and perhaps lack of observer skills (Karl and Steurer 1990; Plantico et al. 1990). However, observations since

1950 have become increasingly standardized or open to correction, and at least those from land are now generally accepted as reliable.

Post-1950 cloud cover trends over land show marked regional differences, but also a common underlying theme. In the USA (Plantico et al. 1990; Sun and Groisman 2004), much of continental Europe (Henderson-Sellers 1992) and more tentatively Australia (P.A. Jones and Henderson-Sellers 1992), cloudiness increased from 1950 to about 1980, then started to fall. In north and south Italy, on the other hand, cloudiness remained steady from 1950 to about 1965, then declined continuously to 1995 (Maugeri et al. 2001). In China (Kaiser 2000) cloud cover declined from 1950 to 1995: but mostly as a sudden fall in 1978, which coincided with an equally abrupt country-wide rise in atmospheric pressures. Kaiser pointed out the concurrence of this with a fundamental switch in the El Niño Southern Oscillation, to greater frequency of El Niño episodes.

Over oceans, Parungo et al. (1994) document a steady rise in cloud cover between 1950 and the early 1980s, as does Norris (1999). The latter record, interestingly, shows a continuing and apparently accelerating rise between 1985 and 1995, in contrast to concurrent trends over land and to satellite observations of global trends between 40°N and 40°S (Rossow and Schiffer 1999). Norris' purely marine data covered 60°N to 40°S, and the recorded increase was nearly all specifically in low cloud, so the two sets are not directly comparable.

The common theme of the ocean and part of the land record is of increasing cloudiness between 1950 and 1980. Thereafter cloudiness falls in all land records. Where there is fall across the full half century, the change is only slight at first but more rapid from about the 1970s.

A pattern seems here to be emerging. Between about 1950 and the late 1970s there is mostly increasing cloud, falling diurnal temperature range and slightly falling mean temperature. For the rest of the 20th century mean temperatures are rising, and land (but not ocean) cloud cover falling and diurnal range falling but ultimately stabilizing. The tendency of diurnal range to continue falling until very late in the period, despite diminishing cloud, may arguably reflect a direct effect of CO_2 in raising night temperatures.

Widespread reductions over the period in water evaporation from open pans (Peterson et al. 1995; Roderick and Farquar 2002) underline the probably major role of mild daytime conditions associated with falling diurnal range and, at least up to the 1970s, more daytime cloud over land. Also to be considered in this context must be atmospheric humidity, its source primarily from the oceans and heavily vegetated land areas. All these factors focus around the hydrological cycle, including the influence of daytime evaporation on surface temperatures and diurnal range.

12.8.3 Evaporative feedbacks

As described in subsection 12.4.2, most mainstream models have prescribed positive feedbacks (some very large) from cloud changes following greenhouse warming. That is, warming reduces cloud, which increases warming, which further reduces cloud and so on. But theory, modelling and history agree that warming increases global evaporation and rainfall: so why not also the amount of low-level cloud that produces the rain and cools most? And should not more rain produce more surface evaporative cooling? All these things would be independent of, and balanced against, the undoubted greenhouse effect of enhanced atmospheric water vapour.

Few studies have yet been able to integrate these factors. Probably the most comprehensive has been that of Dai et al. (1999). Relevant conclusions for summer and autumn in a well-watered land climate were as follows.

1. Clouds are the main factor controlling diurnal range, primarily by reducing maximum temperatures. In this work they had little net effect on minimum temperatures, because carry-over cooling effects from day to night offset warming at night from trapping outgoing long-wave radiation. Clouds reduced mean temperatures mainly in summer.
2. Absolute air humidity was the factor with greatest effect on mean temperature, its greenhouse effect raising both minimum and (to a smaller extent) maximum temperatures. It had a small negative correlation with diurnal temperature range in summer but not in autumn.
3. Soil moisture significantly reduced both maximum temperatures and diurnal range, and more marginally minimum and mean temperatures.

The net overall indication was that cooling by soil surface evaporation and plant transpiration, together with any cooling by resulting extra cloud, exceeded greenhouse warming from the water vapour produced. That agrees with the model findings of Verdecchia et al. (1994), who estimated that such effects would greatly moderate those of doubled atmospheric CO_2 where adequate water was present for evaporation and transpiration. It also agrees with the various studies of land use describes in subsection 12.3.3, where it is shown that the presence of transpiring crops and tropical forests brings about cooling. Under such circumstances the net temperature feedback from the hydrological cycle must be negative. Logically the same can be expected over the oceans.

On the other hand the opposite is predictable over the dry continental interiors, as discussed, for instance, by Ramanathan et al. (1989) and modelled by Verdecchia et al. (1994). With little rain and progressive soil drying, evaporative and cloud cooling diminish, promoting further heating and drying and so on in a positive feedback system. This contributes to the extreme diurnal and annual temperature ranges in these climates, and indicates that general warming from any cause exacerbates their risk of serious heat and drought. (An exception is on the sub-tropical fringe, where warming extends the range and intensity of the summer monsoons.) The records of extreme drought on the American Great Plains and in Central Asia at the time of the Mediaeval Warm Period, described in subsection 12.1.3, lend support to this prediction.

Obviously these are critical considerations for the future of viticulture and of agriculture generally. They indicate that, whether primary temperature changes are natural or anthropogenic, well-watered coastal and near-coastal regions are better buffered than the dry inland against variation in both rainfall and temperature.

12.9   Direct CO_2 effects on plants

12.9.1 *Growth, phenology and plant structure*
Many studies, together with long practical experience of CO_2 fertilization in greenhouse culture (Mortensen 1987; Wittwer 1990) have shown large growth and yield responses to enhanced CO_2 concentrations up to about 1000 parts per million (ppm). That established, main interest here focuses on the nature of the response and factors that can enhance or limit it.

Does CO_2 concentration directly influence plant phenology? Surprisingly little has been reported on this, but the seemingly clear answer is no. Rogers et al. (1992) and Reddy et al. (1997) measured cotton growth under then-existing and doubled CO_2, and found large increases in dry matter but none in the numbers of mainstem nodes and new leaves, a measure of phenological development. In the latter study, rates of node and leaf appearance responded faithfully to temperature across a wide temperature range, but were identical under the two CO_2 concentrations. This agrees with the experience from grapevines as described here and in Gladstones (1992), where it is shown that vine phenological development can be closely predicted from temperature alone, despite wide variation in other factors controlling growth.

We can therefore reasonably assume that future grapevine phenology will be unaffected by rising atmospheric CO_2 directly, but will continue to respond to temperature as in the past.

It follows that any enhanced dry matter from rising CO_2 ends either in existing phenological units or structures (longer and thicker internodes, thicker leaves, stems and roots); in extra branching and root growth; or in greater fruit weight. A common finding is of strongly increased root growth, which as previously argued could be positive for the expression of terroir.

12.9.2 Dependence on temperature

Greenhouse culture experience (Mortensen 1987) shows that full plant response to enhanced CO_2 depends on increased temperature. An enriched greenhouse atmosphere with 700–900 ppm CO_2 is reckoned to require temperatures 2–4°C higher than with normal atmosphere for best fruit yield and quality.

Acock et al. (1990) showed with muskmelons that CO_2 enrichment at sub-optimal temperatures caused undue carbohydrate accumulation in the leaves, unable to be fully used for growth or yield and causing suppression of further photosynthesis. Optimum higher temperature alleviated this by raising the rates of leaf carbohydrate mobilization, transport away, and utilization: a response that is probably universal in plants.

Other research (Grimmer et al. 1999; Turnbull et al. 2002, 2004) has shown that growth responses to enhanced CO_2 depend primarily on night temperatures. This makes sense, since it is the lowest temperatures that most limit the metabolic processes involved, as I postulate also for ripening. For vines it again underlines the need for relatively warm nights and a narrow diurnal temperature range for best function. This need will grow as atmospheric CO_2 and the potential for carbohydrate synthesis both rise. Section 13.4 will take up possible further implications for vine terroirs.

12.9.3 Other limitations to CO2 response

Plants can fail to respond to raised CO_2 for other reasons besides temperature. Nutrient deficiencies are a common cause.

The major nutrients nitrogen and phosphorus are essential components of the photosynthetic apparatus itself, and one or the other is commonly the primary limiting factor for growth even under past lower CO_2 levels (e.g. Conroy et al. 1990; Campbell and Sage 2006). Extra CO_2 will not remedy this.

Additionally, essential trace elements are widely deficient or marginal in the poor soils that carry much of the world's remaining natural vegetation.

Here an initial growth boost by CO_2 can sometimes precipitate or exacerbate a deficiency by locking up more of a nutrient in permanent plant structures or in enhanced acid soil organic matter. Hungate et al. (2004) give an interesting example of this for the trace element molybdenum, essential for nitrogen fixation by the vegetation's legume component.

In dense forest, already complete light interception can likewise cap potential increases in photosynthesis and growth. See Körner (2006) for a thorough review. It is true that higher CO_2 allows photosynthesis at lower light intensities, but that is only a partial compensation.

One aspect in which plants do appear universally to benefit from rising atmospheric CO_2 is in their efficiency of water use (Downton et al. 1980; Wong 1980; Morison and Gifford 1984). More CO_2 passes into plant leaves through their open stomata for the same amount of water lost by transpiration. In practice there is normally a stomatal adjustment allowing a still increased CO_2 intake and photosynthesis but reduced transpiration, giving some increase in growth with a markedly improved ratio of dry matter produced to water transpired. Obviously this is of greatest use in arid regions, where in addition plant growth will tend to be sparse and well illuminated, and soils unleached, thus removing other limitations to growth.

Arid regions aside, though, the capacity of natural vegetation to respond increasing atmospheric CO_2 appears limited. It is different for many forms of agriculture and horticulture, and perhaps especially for viticulture, because for these the limitations to CO_2 response can largely be removed. The following chapter will begin by exploring some direct implications for viticulture.

[CO₂] & altitude

Chapter 13

Current Climate Change and Viticulture

13.1 Direct effects of rising CO$_2$

The responses of greenhouse crops to CO$_2$ levels up to 1,000 ppm shows that they retain at least some of the genetic capacities of their ancestors, most of which evolved in atmospheres much higher in CO$_2$ than exist now.

Hardie (2000) notes that present grapevine (*Vitis*) species evolved in the Cretaceous period, 136–65 million years ago, under atmospheric CO$_2$ concentrations thought to have been around 1,500–3,000 ppm. With such ancestry they could be expected to respond positively to current rises in the same way as other crops. They might also be able to recover adaptation to much higher concentrations through selection or breeding.

The agronomy of cultivated grapevines also means they should respond with fewer of the limitations to which natural vegetation is subject. Proper husbandry counters any limiting nutrient deficits and ensures adequate canopy light exposure. Cropping and pruning, between them, guarantee a low carbon status at the start of the season, giving fullest scope to respond to higher CO$_2$ in both dry matter production and fruitfulness (spring temperatures allowing).

Theoretically, higher CO$_2$ should enable the vine to carry heavier crops while remaining in carbon balance, though this equation could be complex, depending on initial effects on fruitfulness balanced against later capacity to ripen the crop set. Also important in the equation is provision of enough surplus assimilate, after crop needs have been met, for continuing root growth through ripening. If my argument in Chapter 5 is correct, this is essential for the fullest expression of flavour and terroir characteristics in the wine. One immediate future concern is berry nitrogen content. Higher atmospheric CO$_2$ may increase crop and sugar yield, but uptake of nitrogen (which moves readily to the roots) is limited primarily by supply and does not necessarily increase in proportion to extra root growth. Also, photosynthesis continues unabated through fruit ripening at a time when the soil is often drying out and nitrogen uptake impeded. Thus management to ensure adequate berry nitrogen content will probably become more important over time (as perhaps it already has).

It is different for nutrients such as phosphate and most of the trace elements that are largely insoluble and immobile in the soil. Uptake of these is by immediate root-tip contact and chemical exchange with soil minerals, so foraging roots must be dense and their growth continuous for most effective uptake. Stronger root growth under higher CO_2 will in most cases satisfy the plant's correspondingly higher nutrient need, assuming there to be no acute deficiencies.

It also follows that if mineral uptake from deep roots to the berries during ripening plays a key role in terroir expression, as I have argued, then maintaining late deep root growth through moderate cropping and water management will remain as essential as ever for wine quality and terroir expression.

The improving water use efficiency of vines under rising atmospheric CO_2 should improve the vine's stress and growth balance in environments previously too dry and stressful, and thus help extend viticulture into environments with less risk of damaging rains during berry setting and ripening. But as Section 6.4 argues, there are probable limits to how far viticulture can extend into dry atmospheres without compromising quality for table wines.

Rising atmospheric CO_2 concentrations may necessitate some rise in temperature for optimum vine performance. Subsection 12.9.2 noted that greenhouse horticultural crops grown in atmospheres enriched to 700–900 ppm CO_2 need temperatures 2–4°C higher than under normal atmospheres to achieve fully their enhanced yield and quality potentials. Further, the main requirement is for higher night temperatures. Implications for future terroirs will be developed in Section 13.4.

Here we can add some limited evidence on tolerance of temperature extremes. Kriedemann et al. (1976) showed that CO_2 enrichment to 1,200–1,300 ppm substantially improved grapevine survival through heat therapy for virus elimination, using temperatures of 37–40°C day and night for 3–6 months. Enrichment to 950 ppm also enhanced leaf photosynthetic capacity and improved water use efficiency.

Direct evidence relating CO_2 levels to winter cold resistance in grapevines is so far lacking; but given the role of winter-stored carbohydrates in vine cold resistance (Hamman et al. 1996), it can reasonably be inferred that high CO_2 should aid winter (and perhaps spring) survival. However proteins and other compounds also have important roles in freezing resistance, so the issue is complex. Howell (2001) gives a comprehensive review.

Summing up, the evidence shows that rising atmospheric CO_2, in addition to raising vine yield potential, is probably: (a) raising optimum temperatures, especially night temperatures, for vine function; (b) improving tolerance of high temperature extremes; and (c) improving vine drought tolerance and water use efficiency. Effects on cold hardiness are as yet uncertain but could be positive. Some nutrient requirements will probably rise.

13.2 Climate and recent viticultural history

In subsection 12.1.2 I cited studies showing that despite indisputably rising temperatures and advancing vintage dates through the last quarter of the 20th century, best quality vintages in Europe and elsewhere continued mostly to follow the warmer and drier growing seasons of their period, leading to early vintages. This suggested that 20th century growing seasons generally may have been cooler than in the previous centuries when each region's dominant grape varieties became established.

But while that may still be broadly true, a closer examination of the recent record suggests possible alternative reasons for some of the observations.

The first concerns the influence of diurnal temperature range. As subsection 2.1.2 argued, in climates or parts of the growing season cool enough to slow phenological development it is primarily night temperatures that limit. The logic of this, combined with observed vine phenology across climates and mesoclimates with differing diurnal temperature ranges, has made possible a temperature adjustment for diurnal range that significantly improves the fit of vine phenology to standard temperature records. Chapter 11 describes its use in constructing viticultural climate tables.

Based on this adjustment we can estimate the contribution of diurnal range to vine phenology changes over time in the study by G.V. Jones and Davis (2000b) of Bordeaux vintages. These authors showed a progressive and relatively uniform advance of 13 days in harvest date between 1952 and 1997. By contrast the recorded Bordeaux growing season average mean temperatures (G.V. Jones et al. 2005) remained more or less constant between 1952 and 1980 and then rose steeply to 1999.

In my own study of French viticultural climates (Appendix 1), the average growing season diurnal range at Bordeaux airport narrowed by 1.06°C between 1931–60 and 1973–93, while recorded average mean temperature rose by 0.29°C. Part of both changes could have resulted from airport/urban warming. Accepting the figures at face value, and calculations as devel-

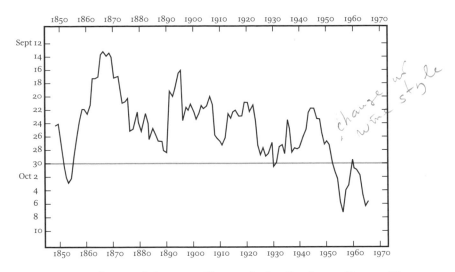

Figure 13.1. Dates of start of vintage at Chateau Lafite, Bordeaux, France. Five-year running averages. From Gladstones (1992), based in data of Ray (1968).

oped in Chapter 11, the resulting predicted advances in harvest date between periods would be seven days, comprising three days due to narrowing diurnal range and four days from rising means. The total predicted advance (allowing for the inexact correspondence of dates) agrees well with the measured trend in maturities as recorded by Jones and Davis.

A puzzling aspect of the Jones and Davis results, which were based on phenological observations and vintage ratings for 10–15 leading quality vineyards, is that across the whole 1952–1997 period the average harvest commencement date was 2 October. Given a total trend to earliness of 'nearly 13 days', that suggests an average date of 8 October at the beginning and 26 September at the end.

Figure 13.1 shows the 5-year running average dates of harvest commencement at the leading Bordeaux vineyard Chateau Lafite, after Gladstones (1992) based on the data of Ray (1968) for the years 1847–1967. The average date between 1952 and 1967 was about 4 October, in good agreement with Jones and Davis' more broadly based estimate for the same period. But as the Lafite figures show clearly, this was a highly atypical period of late vintages, unprecedented in the previous 120 years other than briefly in the mid 1850s when Lafite suffered its initial uncontrolled *Oidium* outbreaks. Nor were late 20th century Bordeaux vintages any earlier than the Lafite average between 1880 and 1950, while Lafite harvests of the 1860s and early 1870s, the years of the great pre-phylloxera vintages, averaged earlier still around the middle and third week of September. Thus even allowing for later picking in recent years because of better fungal disease control, it is hard to argue that the advance in Bordeaux vintages over the second half of the 20th century reflects any climate warming beyond that due to natural fluctuation. This is especially so in that nearly half the maturity advance can be ascribed to falling diurnal temperature range as opposed to rising means.

The question remains as to how much of the progressive improvements in rated wine quality recorded by G.V. Jones and Davis (2000b) for Bordeaux can be attributed to growing season warming, and how much to other factors. At least two such factors, other than management, seem likely on grounds developed in this book.

The first is diminishing diurnal temperature range in its own right, and the second perhaps associated greater cloudiness. Whereas globally the downward trend in diurnal range appears largely to have ceased about 1990 (see subsection 12.8.1), the data of Appendix 1 suggest that in France it continued and accelerated until then or later. The generally inverse correlation between diurnal range and cloudiness would suggest that cloudiness continued to increase over the same period. The Bordeaux vintage ratings show a concentration of outstanding vintages through the 1980s and beginning of the 1990s. Speculatively, then, these seasons combined warmth with temperature equability and cloudiness, but not excessive rain: conditions I have argued produce the best wines.

Nemani et al. (2001) describe a comparable association in California. Parallel 1951–1997 warming in the near-coastal Sonoma and Napa Valley wine-producing regions and the adjacent Pacific Ocean was expressed on land almost entirely by a 2.06°C rise in the annual average minimum temperature, with a 1.13°C rise in the mean and a fall of 1.87°C in the diurnal range. The growing season Temperature Variable Index of Gladstones (1992) fell markedly, from 36.1 to 31.4. Night-time cloudiness and dewpoint on land both increased. The authors attribute these changes to greater evaporation from the warming Pacific and its influence on the land down-wind. They suggest that at least some of the marked improvement in rated wine quality over the period stemmed from these factors.

Neither of these cases proves an association of temperature equability, cloud and humidity with vintage quality, but each provides supporting evidence.

The recent paper of G.V. Jones and Goodrich (2008) adds important evidence on the causes of late 20th century Californian warming. They found little direct relation with phases of the El Niño Southern Oscillation (ENSO), but a strong one with the Pacific Decadal Oscillation (PDO), which they describe as follows.

The PDO is related to the low-frequency variability in sea surface temperatures (SSTs) in the extratropical Pacific Ocean (Frauenfeld et al. 2005) with a
period of around 50 years … Positive (warm phase) values of the PDO refer
to above normal SSTs along the west coast of North America and along the
equator and below normal SSTs in the central and western North Pacific
centred around 45°N latitude …

Jones and Goodrich also cite the view of Mantua and Hare (2002) that there
have been only two complete PDO cycles since 1890, with warm phases spanning 1925–46 and 1977–98 (the end of this phase being still uncertain), and
cold phases 1890–1924 and 1947–76.

An abrupt transition from cold to warm phase in 1976–77 shows clearly
in some of Jones and Goodrich's Californian data, most notably date of last
spring frost. It suggests that much of the Californian temperature rise over
the second half of the 20th century stems from the natural PDO pattern as
opposed to anthropogenic causes or other underlying trends. The longer-
term pattern of PDO phases appears also to correspond well with that of
recorded global temperatures (see Figure 12.6), while at least the later PDO
warm phase was also one of El Niño dominance. Such a combination,
whatever the relationship between the two phenomena, would appear to
provide a strong impetus for the Californian and probably global warming
of the late 20th century. And as G.V. Jones and Goodrich (2008) warn of
Californian viticulture, it foreshadows a possible period of colder and more
erratic vintages once the PDO reverts to its cold phase, lasting up to several
decades or until greenhouse warming catches up.

The same paper demonstrates several further important points. Analysis
of the seasonal factors associated with best vintage quality showed a highly
significant association with lack of spring and ripening-period frosts and
with high average minimum temperature for the growing season. High
growing-season and ripening-period maximum and mean temperatures
contributed independently to high vintage quality. Bloom-time, growing-
season and ripening period rain all had negative effects, but interestingly,
high rainfall during the preceding winter (November–March) contributed
positively to vintage quality. This last finding parallels that of Ashenfelter
et al. (1995) in Bordeaux. It confirms the importance of a thorough winter
wetting of the soil profile, giving greatest capacity for the vine to meet its
moisture needs through ripening from deep soil reserves.

Finally, along with the data of G.V. Jones (2005), it shows that warming
behaviour such as that of the Sonoma and Napa valleys, which in these

Conten-ientaelty look at data

results and those of Nemani et al. (2001) was confined almost wholly to the minimum temperatures, is a property strictly of climates under strong coastal influence. Climates further inland or occluded from the coast appeared to warm, on average, nearly as much in maximum as in minimum temperatures.

We come now to the second factor which might, more speculatively, have contributed to higher vintage quality in the late 20th century: the relationship of atmospheric CO_2 concentration to optimum temperatures for viticulture. May we be seeing the first sign that higher CO_2 raises optimum temperatures, especially optimum minima, for both the vine and wine quality? Only time will answer that question, but the long experience of CO_2-fertilized greenhouse crops (subsection 12.9.2) is suggestive.

A final study of recent viticultural climates requiring scrutiny is that of Petrie and Sadras (2008), which follows the phenology of three grape varieties (Chardonnay, Cabernet Sauvignon and Shiraz), as measured by date of reaching 21.8° Brix sugar content, date of harvest, and °Brix at harvest, in commercial cultivation across 18 regions of south-eastern Australia from 1993 to 2006. It raises some important questions of interpretation.

Using date of reaching 21.8° Brix as the main maturity measure, responses to temperature were gauged in two ways. The first was matching each region's average maturity dates for given varieties against the average mean temperatures at their respective reference recording sites, either for November over the experimental years, or for the warmest month from long-term records after Gladstones (1992). The second approach was to match progressive maturity dates to the time series of recorded temperatures through 1993–2006, using regression analysis.

The first approach gave unambiguous results. Based on 1993–2006 November average mean temperatures at the respective reference sites, maturity averaged over the three varieties was earlier by 6.56 ± 0.92 days per 1°C higher temperature. Using long-term average means for the hottest month, the advance averaged 6.27 ± approximately 0.48 days per 1°C higher temperature. The good correspondence of the two figures is unsurprising: with only small regional differences (in this case) in continentality, the comparative temperature averages for any one month give a mostly adequate indication of those for the full season or year. It is also worth noting that these figures are in broad agreement with corresponding values implied by the method used for maturity prediction by Gladstones (1992) and in this book, based on worldwide climate and phenology comparisons.

Petrie and Sadras' second approach is more problematical. The varietally and regionally averaged advance in attaining 21.8° Brix over the 13 years was 18.42 days. Over the same period the rise in average growing season (October–April) mean temperature, estimated visually from their Figure 2, was 0.87°C; while that for October–December, the phenologically most influential part of the growing season because temperatures then are still mostly below response saturation level, was likewise 0.87°C. This implies a maturity advance per 1°C of temperature rise of 21.7 days, i.e. more than three times that derived from the temperature differences across regions. The clear conclusion, contrary to that of Petrie and Sadras, is that some other element(s) must have been driving earlier sugar maturity over time besides that accounted for by regional temperature differences.

Rising atmospheric CO_2 might theoretically have contributed in its own right. However the contribution, if real, could only have been very small over so short a period. More likely, I believe, is the following explanation stemming from current drought and changing viticultural practice.

The 1993–2006 period in south-eastern Australia was viticulturally notable for two things: progressive drying, with intensifying drought from 2002 onwards, and increasing conversion from flood and sprinkler irrigation to under-vine drip irrigation.

Changing from flood and sprinkler to drip irrigation means that inter-row space remains largely dry through summer and autumn in this predominantly winter-rainfall region. Late spring then brings a natural abrupt reduction in local cooling by evaporation, both from the soil surface and as transpiration by still-green ground cover. To the extent of any combined reduction in rainfall and irrigation water use, there is a similar reduction in direct cooling within the vine canopy by its own transpiration. Sparser foliage also allows more direct solar heating within the canopy. All these mechanisms raise effective vine and fruit temperatures and must advance phenological development and especially berry sugar accumulation. Also greater pre-veraison moisture stress, as is now commonly encouraged, will promote root abscisic acid formation and slightly further advance veraison.

Regionally developing dryness has effects beyond the vineyard. Figure 2 of Petrie and Sadras (2008) shows that nearly all the recorded 1993–2006 warming across the study area was in or around summer: I estimate from it an average of 1.08°C from November to April as against only 0.15°C for May to October. The best fitting explanation for this is again reduced summer evaporation, from the soil, from natural vegetation, and from still-green crops and pastures. Consistent with this, by far the steepest recorded warming, 1.95°C over the 13 years, was in November which is the critical time for drying off of annual crops and pastures. (Petrie and Sadras' selection of November alone as the indicator of warming rate and of maturity advance per degree of warming is therefore misleading, although its use for comparisons across regions should not be greatly compromised since their times of drying off would have been largely synchronous.)

Such regional effects impinge not only on vineyards but also on temperatures as recorded mainly in small regional towns. Any watering restrictions in the towns late in the period would have added further to their recorded warming.

All this shows that neither vine phenology nor even official temperature records over the area and period that Petrie and Sadras studied necessarily reflects climatic change. Intermittent drought is a normal feature of the south-eastern Australian climate. The effects of this drought, together with those of changing viticultural practices, can themselves account for much of the recorded 18.4 days maturity advance over the study period, which is in any case an over-estimate since it is based on sugar content rather than true physiological ripeness.

If instead we accept the more firmly established cross-regional comparisons of phenology and temperature as the basis for calculation, then the recorded average temperature rise from 1993 to 2006 should have advanced sugar ripeness by 5–6 days. But as we have just seen, advance in sugar ripeness still probably gives an over-estimate. The logical conclusion is that the advancing vine phenology that Petrie and Sadras recorded, as in the two other studies just described, does not provide clear evidence of climate change beyond that of known natural fluctuations.

13.3   Projecting future climates

The following discussion focuses speculatively on the decades up to 2050, as being a minimum period for useful planning of new vineyards. Beyond that becomes still more speculative and will receive only brief comment.

13.3.1 Mean temperatures

Records (if flawed) exist of mean temperatures from direct thermometer measurement for the past 150 years, and more approximately from proxy evidence for some hundreds of years. A first approach to projecting future temperatures, then, is to see if any semblance of regularity exists in past fluctuations that could continue into the future.

Some quasi-regularity can indeed by discerned over the past two centuries, with temperature fluctuations showing a periodicity of around 70 years from peak to peak or trough to trough. Thus warming from the Dalton Minimum of the early 19th century led to relative warmth in the 1860s and 1870s, followed by a trough centred around 1900–1910, a new peak about 1940 and a (mild) trough ending in the early to mid 1970s. Warming that started suddenly in 1976 continued steeply to the turn of the millennium, after which land-recorded temperatures stayed more or less constant to 2006. This last warm period appears to have been associated with both a warm phase of the Pacific Decadal Oscillation and a preponderance of El Niño events, and also with a solar maximum as described in subsection 12.2.3. It remains to be seen whether the sharp 2007–2008 cooling, which accompanied a La Niña, is the start of renewed long-term cooling. Solanki et al. (2004) suggest that, having lasted an unusually long 65 years to that date, the 20th century solar maximum was unlikely to last much longer. In any event the pattern of the last 200 years points to a possible cool phase starting soon and potentially lasting to mid century.

Against such projections must be balanced any underlying, and by its nature progressive, warming by greenhouse gases. As argued in Chapter 12 this is unlikely to be more than 0.4–0.5°C for an effective doubling of concentrations. It could be less. A tentative conclusion, then, is that global mean temperatures and their variation up to the mid 21st century will not differ much from those of the late 20th century. Beyond mid 21st century a continuation of the same patterns could see renewed modest warming starting about then and continuing to the end of the century.

13.3.2 Diurnal temperature range and cloudiness
As already discussed, land records of diurnal range show apparently world-wide narrowing from about 1940 to 1990, with cloud increasing in most areas in the earlier part of the period but variably starting to fall again in the later years. Thus the global cooling period from 1940 to 1975 had a predominantly narrowing diurnal range and more land cloud, which seems consistent with cooling from a phase of elevated cosmic ray penetration of the atmosphere (see subsection 12.5.3). By contrast the warming period from 1976 onwards had mostly diminishing land cloud, possibly increasing marine cloud, and an indefinite response of diurnal range. These gross observations give little ground for predicting future trends. However, they do contain a hint that rising CO_2 may be contributing to reduced diurnal range independently of cloud. That accords with its theoretical mode of action. If so we can expect a continuing (probably weak) tendency for the world average of land diurnal range to fall as atmospheric CO_2 rises further, with potentially beneficial effects for viticulture.

More immediately important for viticulture are the very large differences in cloud, humidity and diurnal range existing at any one time across the land, and their potential reactions to overall climate changes. With any general warming the arid mid-latitude continental interiors should become disproportionately hotter and drier in summer, with less cloud, lower humidities and probably wider diurnal temperature range.

The near-coastal Californian viticultural regions, as studied by Nemani et al. (2001) and G.V. Jones and Goodrich (2008), illustrate the opposite response in well-watered, marine-influenced regions. Warming here has significantly increased only the minimum temperatures, with reduced diurnal range and temperature variability index, and increased relative humidities, dew points and night-time cloud. The evident result has been improved wine quality, as was also the case in the similarly maritime climate of Bordeaux (G.V. Jones and Davis 2000b).

Such locations with moderate to warm growing seasons are also the best buffered against falling global temperatures. With climates prone to change in either direction, they offer the safest long-term choice for viticulture. The benefit is strongest for west-facing coasts and their hinterlands, because weather systems move predominantly from west to east and so carry marine influences farthest inland.

13.4   Implications for vine terroirs

Recent concerns for future viticulture and its terroirs have focused almost entirely on climate change and, in particular, greenhouse warming. That would be justified were the more alarmist of greenhouse predictions to prove true. Substantial migrations of viticultural regions would then be needed, or of grape varieties and wine styles within them. But critical examination shows the claims of greenhouse warming to have been poorly based and almost certainly much exaggerated. As discussed in subsection 13.3.1 above, it remains a reasonable expectation that some greenhouse warming will occur through the 21st century, but hardly enough to raise temperatures materially above those of recent decades until after mid century.

Also relevant is that much of any greenhouse warming will stem from rising atmospheric CO_2, which long commercial greenhouse experience suggests will raise optimum minimum temperatures for plant growth and metabolism, and possibly fruit quality. The earlier-noted findings of Ashenfelter et al. (1995) and G.V. Jones and Davis (2000b) in Bordeaux, Grifoni et al. (2006) in Italy, and Nemani et al. (2001) and G.V. Jones and Goodrich (2008) in California seem consistent with this. A logical implication is that superior terroirs will, even more than in the past, require relatively warm nights and warm soil, such as close to coasts, on slopes with good night air drainage, and having aspects and textures that lead to maximum soil warming.

Short-term temperature variability patterns of other types need also to be considered, since they often play a big role in the climate contribution to terroir. Rind et al. (1989) argue logically that temperatures under greenhouse warming will become more equable, not only with less variation between night and day, but also from day to day, month to month and year to year. This stems partly from a shallower tropics to polar temperature gradient, due to most warming being at high latitudes. Air masses that migrate across latitudes to cause such variability will contrast less in temperature and migrate less vigorously. Thus a given rise in temperature means will be less expressed in the extremes, which I argue is favourable to viticulture and wine quality.

Terroir expression should be minimally affected, given that it is remarkably robust across short-term climate fluctuations that occur naturally from season to season and decade to decade. That is how terroir came to be recognized and defined, demonstrating how much of it depends on unchanging local features of geography, topography, soil and underlying geology.

Any warming will certainly allow spread of viticulture polewards and to higher altitudes, and such vineyards may be needed to maintain some cool-climate wine styles in their purest forms. These vines will, however, be least able to exploit the higher CO_2 available to them. Hot and dry inland viticultural areas, through suffering disproportionate heating and drying, may compensate through the vines' greater water use efficiency and heat tolerance. But effects there on wine quality will probably be adverse.

A practical conclusion is that past experience will continue to provide a valid guide to viticultural climates and terroirs for at least some decades to come. Climate tables based on the mid to late 20th century, and calibrated against observed vine phenology for that period, will remain directly relevant.

Appendix 2 gives a limited reference set of completed new tables for representative terroirs in several of the world's leading viticultural regions. These form templates against which further tables can be matched. A more comprehensive set, covering most of the world's viticulture, is currently in preparation and will, I hope, become available soon.

Chapter 14

21st Century Viticulture and its Terroirs: A Summary

14.1 The market

The world of wine in the early 21st century is already very different from that of 50 years ago. Two major changes have been:

1. The industry has become global. Increasingly wine is being traded internationally, with consumption falling in the traditional producing countries but rising elsewhere. Wines from temperate climates of the Southern Hemisphere (Australia, New Zealand, South Africa, Chile, Argentina) have become a sizable component of international trade. New markets are opening up in South and East Asia.

2. Quality standards have risen. Partly this has stemmed from improved technology in the vineyard and winery, and also from selection of better terroirs and grape varieties and retirement of inferior ones. But as well it has been strongly market-driven, as consumers, having experienced better wines, demand still better ones at affordable prices. Concurrently the demand has grown for expensive élite wines.

Both trends will continue. The great terroirs of the Old World will maintain their market prestige and in many cases build on it. But they will be increasingly challenged as the New World identifies equally good terroirs and as its vineyards mature.

Southern Hemisphere countries will exploit their natural advantages for producing high-quality but affordably priced wines on the scales needed for mainstream international trade. The advantages include generally reliable climates; available and largely still-inexpensive land; and growing ability to identify in advance their superior terroirs and suitable grape varieties for them. Regional specialization in locally proven grape varieties and wine styles will add diversity to the mainstream market: witness the outstanding success of Sauvignon Blanc wines from Marlborough, New Zealand. Nor will such new developments necessarily be confined to the Southern Hemisphere. Scope certainly exists in southern and eastern Europe to resurrect and exploit native grape varieties for distinctive wine styles.

Little commercial future exists for the old-style *vins ordinaires* of southern Europe, or for the mass-produced cheap cask or jug wines from hot inland irrigation areas such as of south-eastern Australia and California. As markets become more discriminating they are turning away from these regardless of their cheapness, in favour of wines with freshness and defined characteristics of variety and regional origin: wines that are themselves becoming more price-competitive as terroirs suited to their production are identified and exploited. This does not mean that inferior wines will cease to exist. Nor will it prevent enthusiasts from trying to grow grapes and make wine at the extreme viticultural limits; their produce will undoubtedly still find a place in tolerant local markets. But for wines in national and international commerce the quality bar will continue to rise. Today's mid-price wines will be tomorrow's everyday wines.

14.2   The climate

Chapter 12 examined in some detail the evidence on natural and anthropogenic (man-caused) climate change, and Chapter 13 that of climate in recent decades and its relations to viticulture.

The main conclusion is that warming by anthropogenic greenhouse gases has been much over-estimated. The widely publicized claims of the Intergovernmental Panel on Climate Change (IPCC) and other greenhouse proponents have depended too much on computer models unable to encompass the complexity of real climates; on uncertain data, dubious assumptions and in some key cases biased statistical procedures; and particularly on ignoring the historical record of past climate warmth. Much of the thermometer record of warming over the last 100–150 years, which the IPCC ascribes more or less exclusively to greenhouse gases, has more likely other causes.

Of the estimated rise in *recorded* temperatures over the 20th century of 0.6°C, about half is almost certainly spurious, caused by urban warming around thermometers and an increasing proportion of topographically warmer recording sites. The methods claimed to have removed these biases from the record have been seriously inadequate.

The 20th century's true warming, as recorded in sea surface temperatures, is at least largely accounted for by natural climate fluctuations, for which the most credible cause on decadal to centennial timescales is fluctuation in solar output and magnetic field.

That rising anthropogenic greenhouse gases should make some contribution is theoretically to be expected. The question is how much, with great uncertainties both as to initial effects and to negative feedbacks that, as is now clear, must greatly moderate them. According to mainstream modelling, anthropogenic greenhouse gases should have increased global temperature over the 20th century by 1°C or more. Patently that has not happened. Properly taking into account biases in the land thermometer record and the impact of natural temperature fluctuations, the best conclusion is that greenhouse gases can have caused no more than 0.2°C of warming, which equates to only 0.4–0.5°C temperature rise for a each successive doubling of atmospheric CO_2 or its combined greenhouse equivalent.

The semi-regular temperature fluctuations of the last two centuries suggest further that the natural warming of the late 20th century is at or close to ending, with a likelihood of natural cooling in the near future that will offset any greenhouse warming for some decades to come.

Rising atmospheric CO_2 concentration will itself probably increase the optimum minimum and mean temperatures for vines. This might already have been happening. Thus the possibility cannot be ruled out that the best terroirs will continue their historical shift to warmer locations.

I conclude that the widely held expectation of a viticultural flight to existing cold areas is misplaced. Optimum locations for particular wine styles will probably change little over the coming half century. Any minor shifts will be into areas with higher actual or effective night temperatures, e.g. closer to coasts, to warmer soils or to slopes with superior night air drainage. Favoured sites in warm to hot climates will have higher afternoon relative humidities, most desirably from afternoon sea or lake breezes.

14.3   The terroirs
It follows that terroir studies should again focus on established regional climates and those local factors of geography, topography, mesoclimate, soil, geology and vine nutritional and water relations that together have always constituted terroir as properly understood. The following is a summary of them.

14.3.1 *The climate component*
Quality viticulture can extend over a wide range of growing season and

ripening period mean temperatures, using adapted varieties of varying maturities (see Table 10.1) to fit the available warmth and length of growing season. As a rule of thumb the best-fitting varieties in any climate reach maturity for winemaking in the second half of the first month of autumn (September, March) or just after.

Good viticultural climates have at most moderate temperature variability within the season, i.e. between day and night, day to day and week to week. This minimizes risks from both killing frosts in winter and spring, and damaging high temperatures during veraison and ripening. Relatively warm nights and mild afternoons during ripening promote night synthesis of flavour components while minimizing their daytime loss by evaporation or heat degradation.

Delicate table wines require cloud or low-intensity sunshine (as at high latitudes), fairly high afternoon relative humidities, and no more than moderate temperatures during ripening. They are mostly made from early-maturing grape varieties that just ripen in cool climates. Full-bodied table wines, from predominately midseason grape varieties, need more warmth and sunshine but still at least moderate afternoon relative humidities. Fortified wines need still more warmth and sunshine, and are the most tolerant of ripening-period temperature variability, heat and low relative humidities. All grape varieties and wine styles do best with largely rain-free flowering and ripening periods.

In the absence of irrigation, total rainfall and its distribution should ideally just meet the vines' seasonal needs for health and moderate growth. Ample winter-spring rainfall that fully replenishes soil moisture reserves is always desirable, followed by a dry summer and ripening period that produces enough stress to halt vegetative growth soon after fruit set and then maintain just moderate stress to promote ripening but not leaf loss. Renewed rain between harvest and leaf fall is then desirable to help build up assimilate reserves for the following season. Complementary irrigation, where needed, should aim to duplicate this pattern as closely as possible. It should not be so much as to compromise the vine's natural development and substantial dependence on a deep and extensive root system.

14.3.2 *Geography*

Within regions, much of viticultural climate still depends on local geography.

Proximity to oceans or large lakes governs daily land/sea or lake breeze convection cycles. Afternoon sea and lake breezes are especially important in climates with hot, dry summers, where their early or mid-afternoon arrival greatly mitigates stress on west-facing fruit that become sun-exposed at the hottest time of day. Even later arrival reduces cumulative cluster heating. The same air circulations warm the early mornings, reduce frosts and reduce temperature variability generally. Peninsular and island climates such as those of Italy, New Zealand and Tasmania share all these advantages to varying degrees.

Shelter from arctic winds is essential for vine winter survival in much of North America and Eastern Europe. The Rockies in the west and the Allegheny and Appalachian ranges in the eastern USA help to protect their respective coastal regions by funnelling arctic air masses down the central plains. In south-eastern Europe the Tatra, Carpathian, Transylvanian and Balkan ranges give some protection on more local scales.

14.3.3 Local topography

Whereas broad topography controls regional climates, or macroclimates, local topography and soil characteristics (see subsection 14.3.4 below) provide the variant mesoclimates that help shape individual terroirs.

The best terroirs for quality viticulture are on slopes, particularly of isolated and projecting hills that have unimpeded drainage away of cold air and surplus water. Coastal viticulture, provided the soils are well drained, tolerates flat sites best because mostly their temperatures are already equable.

At high latitudes aspect and steepness of slope both strongly influence the sunlight angle to the soil early and late in the season, and hence warming of the root system at critical times. Steep slopes facing the midday sun have most advantage for early growth and probably flavour ripening, both of which I suggest depend on root-produced hormones favoured by warm and well-drained soils.

A bias towards easterly aspect is favourable because it gives warming in the early morning when it is most needed. Westerly aspects heat most at the hottest time of day, resulting in greatest risk of heat injury in warm to hot climates unless mitigated by reliable sea breezes.

14.3.4 Soil type

Good drainage is essential, with subsoil colour a useful indicator. Yellow, brown and red colours show good drainage. Rocky and stony soils are nearly always well drained. They also typically have only low to moderate fertility, necessary in areas of ample rainfall for controlling vine vegetative vigour.

Depth of root-accessible soil, and that to underlying parent rock if present, are both component dimensions of terroir; but needs vary with climate, with deeper and more water-retentive soils needed in climates having hot, dry summers. Nevertheless in all climates dependence mainly on the soil's deeper water reserves during ripening is positive for wine quality. It gives greatest independence of both drought and unwanted surface wetting; reliable moderate root water stress for production of ripening hormones; and because mineral uptake to the ripening fruit then reflects most closely a site's underlying and unchanging geology, a clearest expression of terroir. See subsection 14.3.5 below.

These ideals depend in turn on subsoil conditions being hospitable for root growth and function, which calls not only for excellent soil drainage and aeration, but also good absorption and transmission of warmth to depth, such as a rocky or stony texture facilitates. The effects complement those directly on the ripening fruit through greater surface absorption of heat during the day and its re-radiation to the vines and fruit at night.

14.3.5 Geology

Two pieces of widely reported circumstantial evidence point to the importance of underlying geology for terroir. First, as noted above, wine terroir characteristics show best where water relations force the vines to depend on their deepest roots for moisture during ripening. Second, they develop fully only once the vines are fully mature, and their roots have reached their greatest depth. Profiles of trace mineral uptake during ripening that reflect underlying geology provide the most plausible explanation.

Geological differences are useful for terroir definition mainly on a local scale, where individual geologies can be most uniform but mutually contrasting, and where climates differ little. Detailed studies of local geology such as those of Meinert and Busacca (2000, 2002) in Washington State, USA, and of Swinchatt and Howell (2004) in the Napa Valley of California, will in time add much to the selection of terroirs in both old and new viticultural areas.

14.3.6 Management influences

While management does not form part of terroir, some of its practices can enhance or mar terroir expression. Some likely to play increasing roles in 21st century viticultural improvement are as follows.

- Minimal cultivation and avoidance of heavy wheeled machinery in the vineyard, to prevent soil hardpan formation and facilitate deep root penetration.

- Better targeted methods of irrigation where this is needed. Note that whereas major dependence on irrigation is probably confined to the cheaper end of the market, premium viticulture in summer-dry areas can still benefit from a capacity to mitigate or forestall excessive stresses during ripening.
- Use of integrated pest management (see subsection 9.1.4).
- Adoption of organic soil management methods as far as feasible. This includes use of composts and mulches rather than artificial fertilizer, and deep-rooted inter-row plant cover in appropriate seasons. The aim is to encourage earthworms and other beneficial soil organisms, build up soil organic matter and structure, and establish and maintain channels of free drainage and circulation of air and nutrients to depth. The rituals of biodynamics confer no further advantage (Section 9.2).

All these measures, besides being sound management, help to create an ecologically favourable image of vineyards and wine.

14.4   Final thoughts

I finish with a quote from the introduction to my book *Viticulture and Environment* (1992).

> In an increasingly standardized, mechanized and computerized world, fewer and fewer things remain that are truly individual. Wine is one of them. Despite the high technology of its production, wine remains essentially a natural product, with infinite variation from area to area, maker to maker, grape variety to grape variety, and even bottle to bottle. It has subtleties and a sensual fascination which take it far beyond being a mere alcoholic beverage. It has a rich social and medical history, as Becker (1988) so eloquently recounts. The moderate consumption of table wines, traditionally with meals within the family or in good company, is a pleasant and entirely healthy custom. These are the things to which post-industrial man is instinctively returning, as a remaining link with the natural world, and as an antidote to the barbarity of his mechanistic surroundings.
>
> Quality in wine is an artistic goal in its own right. Like other artistic goals to which humans aspire, it is a civilizing influence. The world needs such influences.

Better understanding of terroir is bringing these desirable aspirations closer to universal reality, while a better understanding of climate, which must surely develop over the next decade or two, will give further confidence to vineyard planning.

Much of the focus must necessarily be on production of superior everyday wines, since these will form the greater part of the market. Despite having better average terroirs than before, many of their vines will still need supplementary irrigation, which will probably attenuate specific terroir characteristics in the wines. But this still allows distinctive combinations of regional and grape varietal characteristics that will add needed variety and interest in their segment of the market.

Nor is there any real conflict with the production of more highly specified terroir wines. These will remain as prominent as ever in Old World viticulture, perhaps more so, while their equivalents will emerge further in the New World as it identifies – from experience or increasingly through advances in site analysis – its very best terroirs for quality. As in the Old World these will largely be select sites within regions also suited to more extensive production.

In the market place the two levels are complementary rather than competitive. Elite wines do much to establish a region's reputation for all wines. At the same time it is acquaintance with its affordable wines, if good, that encourages many consumers to climb the quality scale and seek out the best.

Sustainable production methods and improving quality and reliability across both market segments will help further establish wine as a world beverage of preference and moderation. The 21st century stands to become wine's Golden Age.

Appendix 1

French Viticultural Temperatures
1931-1993

Three published sets of French climate data allow temperature comparisons over the period 1931–1993: the normals for 1931–1960 (Garnier 1967) and 1961–1990 (World Meteorological Organization (WMO) (1998), and data from 1973 to approximately 1993 from the US National Climatic Data Center (NCDC) (1996). Out of these I have selected sites that meet the following criteria: 1) they span and broadly represent France's viticultural regions; 2) they provide comparisons between at least two of the above periods; and 3) all are from regional airports. The last criterion aims to ensure as far as possible that the recording sites are comparably flat and that there is no major urban bias.

Table A1.1 lists the sites from north to south and indicates the available comparisons. Tables A1.2 to A1.4 are those of average minima, maxima and means for individual sites and months, respectively for 1961–90 vs 1931–60, 1973–93 vs 1961–90, and 1973–93 vs 1931–60. Table A1.5 is a summary comparing averages of annual and growing season (April–October) minimum, maximum and mean temperatures for the five sites having data from all three periods, and the respective averages across all sites in Tables A1.2 to A1.4.

Several trends emerge. The first is a small rise in annual average mean temperature over the full study period, of 0.10–0.15°C depending on how measured. That is effectively over four decades, measuring between the mid points of the first and last recording periods. There is some indication of accelerated warming in the final years. This broadly agrees with the data of Abarca-Del-Rio and Mestre (2006) for the whole of France except that it barely registers their steep temperature rise from 1980 onwards.

Growing season (April–October) mean temperatures, on the other hand, show no overall trend. The difference is because nearly all the warming in the annual averages comes from the months December to February.

Despite a lack of trend for the growing season as a whole, some fluctuations are apparent within it which would undoubtedly have influenced vine phenology and ripening. From 1931–60 to 1961–90 the means from March through June fell by an average of 0.4°C, ample to retard early vine development. By contrast July warmed by 0.27°C, August and September changed little, and October warmed by a massive 0.62°C. The overall impression is of latening seasons.

Comparing 1973–93 with 1961–90, December and January continued their warming trend, but February–April were mixed, with a slight net cooling. May through September then showed minor warming, particularly in August, but the outstanding 1961–90 October warmth did not persist. The suggestion is of a continued cool spring but further slight summer warming.

Annual and seasonal trends in the means were largely consistent across regions. However certain stations showed anomalous behaviour. Lyon recorded 0.06°C cooling in annual average between 1931–60 and 1961–90, but 0.29°C warming between 1961–90 and 1973–93. Bordeaux warmed 0.25°C between 1931–60 and 1961–90, and a further 0.11°C between 1961–90 and 1973–93, while Nice warmed 0.28°C between 1931–60 and 1961–90 and a further 0.05°C from 1961–90 to 1973–93 Notably these three are major cities and airports, with the airports close to the cities in the cases of Lyon and Nice. Bordeaux airport is further from the city centre, but Johnson and Robinson (2001) show it as only just beyond its outskirts. All three are therefore likely to have undergone some artificial warming, from either or both of urban expansion and the growth of busy airports and their associated industries.

By contrast Perpignan registered a cooling trend throughout: by 0.20°C for the year and 0.33°C for the growing season between 1931–60 and 1961–90, and a further 0.05°C for the year and 0.15°C for the growing season from 1961–90 to 1973 93. The airport in this case is isolated from the town and perhaps not large or busy enough to generate warming of its own. Possible thermometer shifts to cooler sites must also be considered.

The balance of these anomalies, together with presumed airport growth at other centres, points to a likely small warming bias across the record. The conclusion, then, is of little trend in either annual or growing season temperature averages, but of some country-wide changes in the seasons, with clear warming in winter, cooling in spring in the earlier years, and minor but irregular warming in summer and early autumn. These findings do not preclude an accelerating general warming after the mid 1980s as shown by Abarca-Del-Rio and Mestre (2006).

All the above refers to averages of the temperature means. But looking at the minima and maxima separately, a different picture emerges. Over the study period the two converge by approximately 1.2°C. The minima rise by 0.6°C and the maxima fall by 0.6°C.

Closer inspection shows this to have been an accelerating process. Between 1931–60 and 1961–90, an average gap of thirty years, measured convergence was 0.45°C. But from 1961–90 to 1973–93, an average gap of only eight years, it was nearly 0.7°C. Most rapid convergence must therefore have been from the 1970s on, a time when the means were probably starting to rise.

Section 13.2 of the main text discusses some implications in a wider context.

Table A1.1. France: temperature comparisons between recording periods. Locations north to south.

Location	Airport name	Latitude North	1961–90 vs 1931–60	1973–93 vs 1961–90	1973–93 vs 1931–60
Nancy	Tomblaine	48°41'			✔
Strasbourg	Entzheim	48°33'			✔
Le Mans	Arnage	47°56'			✔
Dijon	Longvic	47°16'	✔		
Nantes	Ch. Bourgon	47°10'	✔	✔	✔
Bourges	-	47°04'	✔		
Lyon	Bron	45°43'	✔	✔	✔
Bordeaux	Mérignac	44°50'	✔	✔	✔
Nîmes	Courbessac	43°52'	✔		
Nice	Côte d'Azur	43°39'	✔	✔	✔
Toulouse	Blagnac	43°38'	✔		
Marseille	Marignane	43°27'		✔	
Perpignan	Llabanère	42°44'	✔	✔	✔

Table A1.2. France: differences in average minima, maxima and means, °C, 1961–90 vs 1931–60.

MIN	Dijon	Nantes	Bourges	Lyon	Bordeaux	Nîmes	Toulouse	Nice	Perpignan	Average
Jan	+0.6	+0.7	+0.5	+0.5	+0.6	+0.6	+1.0	+0.7	+0.4	+0.62
Feb	+1.0	+1.0	+0.9	+1.3	+1.1	+1.2	+1.6	+1.0	+0.5	+1.07
Mar	+0.2	−0.1	+0.1	−0.1	−0.3	−0.1	−0.1	+0.5	−0.6	−0.06
Apr	−0.1	−0.1	−0.2	−0.2	+0.2	0.0	+0.4	+0.7	−0.5	+0.02
May	0.0	+0.3	−0.3	0.0	+0.4	+0.1	+0.2	+0.5	−0.5	+0.08
June	−0.1	+0.2	−0.4	−0.2	+0.2	+0.2	−0.2	+0.4	−0.4	−0.03
July	+0.2	+0.4	−0.1	+0.2	+0.7	+0.5	+0.5	+1.1	−0.2	+0.37
Aug	+0.2	+0.2	−0.2	0.0	+0.6	+0.2	+0.1	+1.3	−0.4	+0.22
Sept	0.0	+0.2	−0.3	−0.2	+0.2	+0.1	−0.2	+0.6	−0.8	−0.04
Oct	+0.8	+0.9	+0.6	+0.9	+0.9	+0.6	+1.1	+1.2	−0.1	+0.77
Nov	−0.2	+0.3	0.0	−0.2	+0.4	+0.1	+0.4	+0.6	−0.3	+0.12
Dec	+0.1	+0.6	+0.2	+0.1	+0.4	0.0	+0.4	+0.7	−0.3	+0.24
Year	+0.225	+0.383	+0.067	+0.175	+0.450	+0.292	+0.433	+0.775	−0.267	+0.281
Apr-Oct	+0.143	+0.300	−0.129	+0.071	+0.457	+0.243	+0.271	+0.829	−0.414	+0.197

MAX										
Jan	−0.2	+0.2	−0.1	+0.3	+0.2	+0.3	+0.5	0.0	+0.3	+0.17
Feb	+0.5	+0.3	+0.4	+0.7	+0.7	+0.1	+0.7	−0.2	+0.3	+0.39
Mar	−1.4	−1.0	−1.6	−1.2	−1.2	−0.6	−1.0	−0.3	−0.7	−1.00
Apr	−1.1	−1.0	−1.2	−1.1	−1.0	−0.7	−0.8	−0.4	−1.0	−0.92
May	−1.1	−1.0	−1.1	−1.0	−0.6	−0.3	−0.3	−0.6	−0.4	−0.71
June	−0.6	−0.7	−1.0	−0.9	−0.5	−0.5	−0.5	−0.6	−0.4	−0.63
July	+0.2	+0.2	+0.3	−0.1	+0.6	−0.1	+0.6	−0.1	0.0	+0.18
Aug	−0.2	−0.5	−0.1	−0.4	−0.1	−0.4	−0.1	0.0	−0.4	−0.24
Sept	0.0	+0.1	0.0	−0.2	+0.6	−0.2	+0.6	−0.6	+0.1	+0.04
Oct	+0.2	+0.3	+0.6	+0.6	+0.8	+0.4	+0.9	+0.1	+0.4	+0.48
Nov	−0.5	+0.2	−0.2	−0.1	+0.4	−0.2	+0.2	−0.3	−0.2	−0.08
Dec	0.0	+0.4	0.0	−0.1	+0.6	+0.3	+0.7	+0.3	+0.3	+0.28
Year	−0.350	−0.208	−0.333	−0.292	+0.042	−0.158	+0.125	−0.225	−0.142	−0.170
Apr-Oct	−0.371	−0.372	−0.357	−0.443	−0.029	−0.257	+0.057	−0.314	−0.243	−0.258

MEAN										
Jan	+0.20	+0.45	+0.20	+0.40	+0.40	+0.45	+0.75	+0.35	+0.35	+0.394
Feb	+0.75	+0.65	+0.65	+1.00	+0.90	+0.65	+1.15	+0.40	+0.40	+0.727
Mar	−0.60	−0.55	−0.75	−0.65	−0.75	−0.35	−0.55	+0.10	−0.65	−0.528
Apr	−0.60	−0.55	−0.70	−0.65	−0.40	−0.35	−0.20	+0.15	−0.75	−0.450
May	−0.55	−0.35	−0.70	−0.50	−0.10	−0.10	−0.05	−0.05	−0.45	−0.317
June	−0.35	−0.25	−0.70	−0.55	−0.15	−0.15	−0.35	−0.10	−0.40	−0.333
July	+0.20	+0.30	+0.10	+0.05	+0.65	+0.20	+0.55	+0.50	−0.10	+0.272
Aug	0.00	−0.15	−0.15	−0.20	+0.25	−0.10	0.00	+0.65	−0.40	−0.010
Sept	0.00	+0.15	−0.15	−0.20	+0.40	−0.05	+0.20	0.00	−0.35	0.000
Oct	+0.50	+0.60	+0.60	+0.75	+0.85	+0.50	+1.00	+0.65	+0.15	+0.622
Nov	−0.35	+0.25	−0.10	−0.15	+0.40	−0.05	+0.30	+0.15	−0.25	+0.022
Dec	+0.05	+0.50	+0.10	0.00	+0.50	+0.15	+0.55	+0.50	0.00	+0.261
Year	−0.063	+0.088	−0.133	−0.058	+0.246	+0.067	+0.279	+0.275	−0.204	+0.055
Apr-Oct	−0.114	−0.036	−0.243	−0.186	+0.214	−0.007	+0.164	+0.257	−0.329	−0.031

Table A1.3. France: differences in average minima, maxima and means, °C, 1973–93 vs 1961–90.

MIN	Nantes	Lyon	Bordeaux	Nice	Marseille	Perpignan	Average
Jan	+0.32	+0.79	+0.09	+0.43	+0.13	+0.34	+0.350
Feb	−0.08	+0.28	+0.01	+0.29	−0.19	+0.33	+0.107
Mar	+0.39	+0.81	+0.54	+0.47	+0.02	+0.78	+0.502
Apr	+0.10	+0.22	+0.09	−0.09	−0.38	+0.17	+0.018
May	+0.44	+0.76	+0.50	+0.39	−0.03	+0.47	+0.422
June	+0.32	+1.07	+0.54	+0.38	+0.10	+0.49	+0.483
July	+0.32	+1.06	+0.49	+0.42	+0.22	+0.62	+0.522
Aug	+0.50	+0.93	+0.69	+0.76	+0.28	+0.70	+0.643
Sep	+0.14	+0.63	+0.24	+0.43	+0.21	+0.41	+0.343
Oct	−0.12	+0.41	+0.01	−0.09	−0.32	−0.06	−0.028
Nov	+0.01	+0.32	+0.07	+0.12	−0.36	+0.29	+0.075
Dec	+0.61	+1.21	+0.60	+0.41	+0.21	+0.54	+0.597
Year	+0.246	+0.708	+0.323	+0.327	−0.009	+0.423	+0.336
Apr-Oct	+0.243	+0.726	+0.366	+0.314	+0.011	+0.400	+0.343

MAX							
Jan	−0.01	−0.03	−0.01	0.00	−0.31	−0.21	−0.095
Feb	−0.27	−0.54	−0.14	−0.32	−0.43	−0.53	−0.372
Mar	−0.03	+0.23	+0.13	0.00	−0.53	−0.31	−0.075
Apr	−0.51	−0.59	−0.36	−0.59	−0.80	−0.87	−0.620
May	+0.08	−0.08	−0.03	−0.30	−0.84	−0.96	−0.355
June	−0.29	−0.31	−0.31	−0.26	−0.19	−0.89	−0.375
July	−0.29	−0.06	−0.17	−0.29	−0.28	−0.63	−0.287
Aug	+0.33	+0.46	+0.40	+0.12	−0.07	−0.22	+0.170
Sept	−0.24	−0.23	−0.14	−0.09	+0.02	−0.47	−0.192
Oct	−0.86	−0.80	−0.84	−0.52	−0.61	−0.87	−0.750
Nov	−0.27	−0.27	−0.04	−0.31	0.27	−0.38	−0.257
Dec	+0.28	+0.60	+0.32	−0.06	+0.30	+0.03	+0.245
Year	−0.173	−0.135	−0.099	−0.218	−0.334	−0.526	−0.247
Apr-Oct	−0.254	−0.230	−0.207	−0.276	−0.396	−0.701	−0.344

MEAN							
Jan	+0.155	+0.380	+0.040	+0.215	−0.090	+0.065	+0.128
Feb	−0.175	−0.130	−0.065	−0.015	−0.310	−0.100	−0.133
Mar	+0.180	+0.520	+0.335	+0.235	−0.255	+0.235	+0.208
Apr	−0.205	−0.185	−0.135	−0.340	−0.590	−0.350	−0.301
May	+0.260	+0.340	+0.235	+0.045	−0.435	−0.245	+0.033
June	+0.015	+0.380	+0.115	+0.060	−0.045	−0.200	+0.054
July	+0.015	+0.500	+0.160	+0.065	−0.030	−0.005	+0.118
Aug	+0.415	+0.695	+0.545	+0.440	+0.105	+0.240	+0.407
Sept	−0.050	+0.200	+0.050	+0.170	+0.115	−0.030	+0.076
Oct	−0.490	−0.195	−0.415	−0.305	−0.465	−0.465	−0.389
Nov	−0.130	+0.025	+0.015	−0.095	−0.315	−0.045	−0.091
Dec	+0.445	+0.905	+0.460	+0.175	+0.255	+0.285	+0.421
Year	+0.036	+0.286	+0.112	+0.054	−0.172	−0.052	+0.044
Apr-Oct	−0.006	+0.248	+0.079	+0.019	−0.192	−0.151	0.000

Table A1.4. France: differences in average minima, maxima and means, °C, 1973–93 vs 1931–60.

MIN	Nancy	Strasbourg	Le Mans	Nantes	Lyon	Bordeaux	Nice	Perpignan	Average
Jan	+1.13	+1.30	+1.18	+1.02	+1.29	+0.69	+1.13	+0.74	+1.060
Feb	+0.48	+0.81	+1.09	+0.92	+1.58	+1.11	+1.29	+0.83	+1.014
Mar	+1.28	+1.17	+1.23	+0.29	+0.71	+0.24	+0.97	+0.18	+0.759
Apr	−0.19	−0.33	+0.48	0.00	+0.02	+0.29	+0.61	−0.33	+0.069
May	+0.48	+0.41	+0.79	+0.74	+0.76	+0.90	+0.89	−0.03	+0.618
June	+0.48	+0.56	+0.76	+0.52	+0.87	+0.74	+0.78	+0.09	+0.600
July	+0.46	+0.32	+1.22	+0.72	+1.26	+1.19	+1.52	+0.42	+0.889
Aug	+0.57	+0.28	+1.12	+0.70	+0.93	+1.29	+2.06	+0.30	+0.906
Sept	+0.13	+0.20	+0.39	+0.34	+0.43	+0.44	+1.03	−0.39	+0.321
Oct	+0.77	+0.86	+1.17	+0.78	+1.13	+0.91	+1.11	−0.16	+0.844
Nov	−0.19	−0.24	+0.38	+0.31	+0.12	+0.47	+0.72	−0.01	+0.195
Dec	+1.16	+1.22	+1.23	+1.21	+1.31	+1.00	+1.11	+0.24	+1.060
Year	+0.547	+0.545	+0.920	+0.629	+0.883	+0.773	+1.102	+0.157	+0.695
Apr-Oct	+0.386	+0.333	+0.847	+0.543	+0.797	+0.823	+1.143	−0.014	+0.607

MAX	Nancy	Strasbourg	Le Mans	Nantes	Lyon	Bordeaux	Nice	Perpignan	Average
Jan	+0.41	+0.60	−0.06	+0.19	+0.27	+0.19	0.00	+0.09	+0.211
Feb	−0.01	+0.10	−0.24	+0.03	+0.16	+0.56	−0.52	−0.23	−0.019
Mar	−0.83	−0.49	−0.99	−1.03	−0.97	−1.07	−0.30	−1.01	−0.836
Apr	−1.47	−1.50	−1.66	−1.51	−1.69	−1.36	−0.99	−1.87	−1.506
May	−0.94	−0.81	−1.36	−0.92	−1.08	−0.63	−0.90	−1.36	−1.000
June	−1.11	−1.00	−1.64	−0.99	−1.21	−0.81	−0.84	−1.29	−1.111
July	−0.09	−0.30	−0.54	−0.09	−0.16	+0.43	−0.39	−0.63	−0.221
Aug	+0.07	+0.13	−0.52	−0.17	+0.06	+0.30	+0.12	−0.62	−0.079
Sept	−0.51	−0.08	−0.69	−0.14	−0.43	+0.46	−0.69	−0.37	−0.306
Oct	−0.61	−0.07	−0.27	−0.56	−0.20	−0.04	−0.42	−0.47	−0.330
Nov	−0.77	+0.01	−0.36	−0.07	−0.37	+0.36	−0.61	−0.58	−0.299
Dec	+0.78	+1.16	+0.54	+0.68	+0.50	+0.92	+0.24	+0.33	+0.644
Year	−0.423	−0.192	−0.649	−0.382	−0.427	−0.058	−0.442	−0.668	−0.405
Apr-Oct	−0.666	−0.519	−0.954	−0.626	−0.673	−0.236	−0.587	−0.944	−0.650

MEAN	Nancy	Strasbourg	Le Mans	Nantes	Lyon	Bordeaux	Nice	Perpignan	Average
Jan	+0.770	+0.950	+0.560	+0.605	+0.780	+0.440	+0.565	+0.415	+0.636
Feb	+0.235	+0.455	+0.425	+0.475	+0.870	+0.835	+0.385	+0.300	+0.498
Mar	+0.225	+0.340	+0.120	−0.370	−0.130	−0.415	+0.335	−0.415	−0.039
Apr	−0.830	−0.915	−0.590	−0.755	−0.835	−0.535	−0.190	−1.100	−0.719
May	−0.230	−0.200	−0.285	−0.090	−0.160	+0.135	−0.005	−0.695	−0.191
June	−0.315	−0.220	−0.440	−0.235	−0.170	−0.035	−0.030	−0.600	−0.256
July	+0.185	+0.010	+0.340	+0.315	+0.550	+0.810	+0.565	−0.105	+0.334
Aug	+0.320	+0.205	+0.300	+0.265	+0.495	+0.795	+1.090	−0.160	+0.414
Sept	−0.190	+0.060	−0.150	+0.100	0.000	+0.450	+0.170	−0.380	+0.008
Oct	+0.080	+0.395	+0.450	+0.110	+0.555	+0.435	+0.345	−0.315	+0.257
Nov	−0.480	−0.115	+0.010	+0.120	−0.125	+0.415	+0.055	−0.295	−0.052
Dec	+0.970	+1.190	+0.885	+0.945	+0.905	+0.960	+0.675	+0.285	+0.852
Year	+0.062	+0.180	+0.135	+0.124	+0.228	+0.358	+0.330	−0.255	+0.145
Apr-Oct	−0.140	−0.096	−0.054	−0.041	+0.062	+0.294	+0.278	−0.479	−0.022

Table A1.5. Annual and growing season (April–October) average temperatures across viticultural regions of France: summary of differences between recording periods, °C.

	1961–90 vs 1931–60			1973–93 vs 1961–90			1973–93 vs 1931–60		
ANNUAL	Min	Max	Mean	Min	Max	Mean	Min	Max	Mean
Nantes	+0.383	−0.208	+0.088	+0.246	−0.173	+0.036	+0.629	−0.382	+0.124
Lyon	+0.175	−0.292	−0.058	+0.708	−0.135	+0.286	+0.883	−0.427	+0.228
Bordeaux	+0.450	+0.042	+0.246	+0.323	−0.099	+0.112	+0.773	−0.058	+0.358
Nice	+0.775	−0.225	+0.275	+0.327	−0.218	+0.054	+1.102	−0.442	+0.330
Perpignan	−0.267	−0.142	−0.204	+0.423	−0.526	−0.052	+0.157	−0.668	−0.255
Average	+0.303	−0.165	+0.069	+0.405	−0.230	+0.087	+0.709	−0.395	+0.157
Averages from:									
Table 2 (9 sites)	+0.281	−0.170	+0.055						
Table 3 (6 sites)				+0.336	−0.247	+0.044			
Table 4 (8 sites)							+0.695	−0.405	+0.145
APR–OCT									
Nantes	+0.300	−0.372	−0.036	+0.243	−0.254	−0.006	+0.543	−0.626	−0.041
Lyon	+0.071	−0.443	−0.186	+0.726	−0.230	+0.248	+0.797	−0.673	+0.062
Bordeaux	+0.457	−0.029	+0.214	+0.366	−0.207	+0.079	+0.823	−0.236	+0.294
Nice	+0.829	−0.314	+0.257	+0.314	−0.276	+0.019	+1.143	−0.578	+0.278
Perpignan	−0.414	−0.243	−0.329	+0.400	−0.701	−0.151	−0.014	−0.944	−0.479
Average	+0.249	−0.280	−0.016	+0.410	−0.334	+0.038	+0.658	−0.613	+0.023
Averages from:									
Table 2 (9 sites)	+0.197	−0.258	−0.031						
Table 3 (6 sites)				+0.343	−0.344	0.000			
Table 4 (8 sites)							+0.607	−0.650	−0.022

Appendix 2

Reference Climate Tables

Introduction and comments

The 18 tables presented here illustrate the range from cool to hot viticultural climates, and represent mostly well-known wine regions against which new tabulations can be compared. The section Site Adjustments and Notes that precedes them gives examples of data estimates made where direct records were lacking, and of adjustments to data from official recording sites for regionally typical vineyards as discussed in subsection 2.1.5 and Chapter 4 and summarized in Table 2.2.

Each climate table specifies the periods of temperature records used and their sources. Other data, e.g. for sunshine hours, are from the same sources except as noted in the section on site adjustments preceding the tables. (Few sources on their own provide all the records needed for a comprehensive climate description.) As far as possible I have used records from comparable periods of the mid to late 20th century, with the 1961–90 normals as a standard. Not all those available cover the theoretically required span of 30 years, but most are for 20 years or more.

Greatest possible accuracy is essential for monthly averages of minimum and maximum temperatures because on these depend the estimates of ripening dates, and hence the estimates for all climate elements during ripening. Accuracy of the basc data for the other, although important, is less critical. More general indications of these usually suffice for the climate comparisons needed here. If unavailable directly, I have estimated them were possible from closely enough related sites or earlier (but post-1930) recording periods; or in the cases of sunshine hours and cloud, by interpolation from climate maps. As an important instance of estimation, the V_{min} and V_{max} temperature factors that measure variability of the minima and maxima, being closely related to permanent features of the terrain, can be extrapolated with fair safety between periods or across nearby sites of similar terrain. This makes possible a worthwhile estimation of monthly average lowest minima and highest maxima, together with the full range of derived indices, for some of the many sites for which the only available temperature records are of their monthly average minima and maxima.

Six of the 18 tables (for Dijon, Geisenheim, Vienna, Geneva NY, Blenheim and Launceston) describe cool viticultural climates, and are alike in favouring grape varieties of maturity groups 3 and 4 despite the first four being highly continental and the last two highly maritime. They show how maritime influence at relatively low latitude can achieve equability during ripening comparable to that experienced in high-latitude cool climates.

Three sites (Jerez de la Frontera, Fresno and Mildura) can be described as unequivocally hot, but contrast in significant ways. Fresno and Mildura are both inland, arid, and subject to high temperature variability. Fresno has very high average summer maxima and means, with a very wide diurnal range but only limited variability of the maximum (V_{max}). Both the latter features stem from its valley-floor site, protected by surrounding mountains from sources of still hotter and cooler air. Mildura's average maxima and means are lower, with a narrower (but still wide) diurnal range; but variability of the maximum is much greater than that of Fresno, resulting in both having similar average highest maxima. The difference is because Mildura is on a flat plain, directly exposed to periodic extremely hot winds from the central Australian desert (such as seriously affected its 2009 vintage). In contrast to both, and despite its high average temperatures, Jerez de la Frontera is equable because of its coastal location.

The remaining sites are in varying ways intermediate, having in common that all are best known for medium to full-bodied red table wines. But these also have some interesting contrasts that can be related to their respective geographies.

The two west-coastal American sites, Saint Helena (California) and Santiago (Chile) both have notably cold nights and wide diurnal temperature ranges, due to their valley-floor locations. But as with Fresno this is offset, particularly for Santiago, by low variability of the maxima which results in favourably moderate average highest-maxima. Combined with very low ripening-period rainfall, suitable ripening-period mean temperatures and – at least for Santiago – adequate afternoon relative humidities, this gives a generally good climatic regime for table wine production other than for a rather high ends-of-season frost risk.

Mendoza provides an interesting comparison. Its seasonal and ripening-period mean temperatures point to a hot climate. But these are tempered by much greater equability than in the two previous examples, with ripening-period highest maxima appreciably below those of Saint Helena and temperature variability factors substantially less than of either. Combined with greater cloudiness and adequate afternoon relative humidities, this makes possible the production of good, full-bodied red wines, summer rains and hail allowing (see Johnson and Robinson 2001). Much the same can be said of Marseille, typifying the near-coastal regions of Southern France.

A more general conclusion can be justified. The close comparability of adapted grape varieties and of wine styles across environments with similar E° days and ripening conditions validates three things: the use of E° days as a measure of phenologically effective heat for grapevines, the common need for and approximate effectiveness of temperature adjustments for site characteristics, and the cardinal role of final ripening conditions in determining grape and wine characteristics.

One important difference remains between continental (Geisenheim, Vienna, Geneva NY) and maritime (Blenheim, Launceston) cool viticultural climates. The former have advantages for late-picked, sweet styles, which depend on final, very cool and humid conditions for sugar concentration under 'noble rot' influence while retaining aromatics and natural acids; whereas the latter, autumn rains allowing, enjoy a wider range of varietal adaptation and reliability of ripening for dry wines.

Site Adjustments and Notes

Dijon (Burgundy), **France** (Table A2.1)
Vineyard site adjustments, °C

Min.	Max.	
+0.2	+0.2	Vineyards S of Dijon
–0.3	–0.3	Average vineyard altitude c. 275 m
+0.4	–	Moderate slopes, E aspects
+0.4	–	Calcareous soils
+0.7	–0.1	Mean +0.3, range –0.8

Notes
V_{min} and V_{max} from Lyon 1973–93 (NCDC 1996); R.H. ex MOUK (1967).
Principal grape varieties Pinot Noir and Chardonnay (both Maturity
Group 3).

Bordeaux, **France** (Table A2.2)
Vineyard site adjustments, °C

Min.	Max.	
–	–	Vineyards both N and S
+0.1	+0.1	Average vineyard altitude c. 40 m
+0.2	–	Mostly undulating
+0.2	–0.2	Mostly closer to Gironde Estuary
+0.4	–	Stony or calcareous soils
+0.9	–0.1	Mean +0.4, range –1.0

Notes
V_{min} and V_{max} 1973–93 ex NCDC (1996); cloud cover from Garnier (1967).
Principal grape varieties Sauvignon Blanc (Maturity Group 3), Semillon
(4), Merlot (5) and Cabernet Sauvignon (6).

Mérignac Airport is a little west of Bordeaux, and represents most nearly
the Graves appellation. However with the listed site adjustments it should
reasonably represent the Bordeaux region generally.

Marseille, France (Table A2.3)

Vineyard site adjustments, °C

Min.	Max.	
–0.4	–0.4	Typical vineyard altitude 100 m
+0.4	–	Moderate slopes, all aspects
+0.4	–	Stony and calcareous soils
+0.4	–0.4	Mean unchanged, range –0.8

Notes

V_{min}, V_{max} and R.H. 1973–93 ex NCDC (1996); cloud ex Garnier (1967). Prominent grape varieties are Syrah, Cinsaut (both Maturity Group 5), Grenache (7) and Carignan and Mourvèdre (8).

Marignan Airport is somewhat inland from Marseille, and sheltered behind coastal hills as reflected in its lower 4 p.m. R.H. compared with the immediate coast. With adjustments for altitude etc. it should reasonably represent the lower-altitude vineyards of Provence and E Languedoc.

Geisenheim, Germany (Table A2.4)

Vineyard site adjustments, °C

Min.	Max.	
–0.2	–0.2	Average vineyard altitude c. 150 m
+0.4	–	Moderately greater slope
+0.4	–	Mostly S aspects
+0.4	–	Calcareous or gravelly soils
+1.0	–0.2	Mean +0.4, range –1.2

Notes

V_{min}, V_{max} and R.H. as for Frankfurt-am-Main Airport for 1973–90 ex NCDC (1996)

Geisenheim is in the heart of the Rheingau, and the estimates are for its better terroirs where the principal grape variety is Riesling (Maturity Group 4).

Vienna, Austria (Table A2.5)

Vineyard site adjustments, °C

Min.	Max.	
–0.6	–0.6	Average vineyard altitude c. 280 m
+0.4	–	Moderate slopes
+0.4	–	Largely SE aspects
+0.2	–0.6	Mean –0.2, range –0.8

Notes

Sun hours, cloud and rainfall are from Vienna-Hohe Warte, 48°15'N, 16°22'E, alt. 209 m for 1961–90 (WMO 1998).

The estimate is primarily for Vienna's most famous vineyards of the Wachau and Kremstal upstream from the city. Described soil types are varied, attracting no overall temperature adjustment. Leading grape varieties are Gewürztraminer and Pinot Noir (both Maturity Group 3) and Riesling and Grüner Veltliner (4).

Torino, Italy (Table A2.6)
Vineyard site adjustments, °C

Min.	Max.	
−0.1	−0.1	Average vineyard altitude c. 300 m
+0.4	−	Moderate slopes, most aspects
+0.4	−	Gravels and marl soils
+0.7	−0.1	Mean +0.3, range −0.8

Notes

Sun hours from WMO (1998). The adjusted data for Torino broadly represent a swathe of hilly viticultural regions to the east and south-east, producing mostly red wines of which the most famous are Barolo and Barbaresco. Principal grape varieties are Barbera (Maturity Group 6) and Nebbiolo and Grignolino (both 7). The notably late-ripening Nebbiolo demands the warmest exposures, and is described as ripening typically under the local autumn fogs after which it is named.

Jerez de la Frontera, Spain (Table A2.7)
Vineyard site adjustments, °C

Min.	Max.	
−0.2	−0.2	Average vineyard altitude c. 60 m
+0.2	−	Gently undulating
−	−0.8	Sea breezes
+0.8	−	Very calcareous soils
+0.8	−1.0	Mean −0.1, range −1.8

Notes

Sun hours are from WMO (1998), and the main grape variety Palomino (Maturity Group 6), used for sherry making.

Jerez differs from the two other hot viticultural climates illustrated here (Fresno, Table A2.8, and Mildura, Table A2.15) in being near-coastal, resulting in a more equable climate with regular summer sea breezes. This shows in its relatively low heat stress and temperature variability indices, and higher 4 p.m. relative humidities. The highly absorptive calcareous soils also provide a steady water supply to the vines, despite very low summer rainfall. The combination results in bland but sound base wines, well suited to sherry making.

Fresno, California, USA (Table A2.8)
Vineyard site adjustments: none.

Notes

Sun hours and cloud cover are from WMO (1998). Fresno typifies the hot, mid San Joaquin Valley and produces bulk wines from diverse grape varieties under full irrigation. Notable during ripening are its very long sun hours, lack of cloud or rain, and very low relative humidities. The heat is, however, regular, as shown by high average maxima but low values of V_{max}.

Saint Helena (Napa V.), California, USA (Table A2.9)
Vineyard site adjustments: none.

Notes

V_{min} estimated ex NCDC (1996) as for Sacramento Airport –0.3°C, V_{max} the same +0.2°C, in recognition of St Helena's wider diurnal range; sun hours and cloud interpolated from the maps of NCDC (2002). Grape varieties include Chardonnay (Maturity Group 3), Zinfandel and Petite Sirah (? syn. Durif) (4), and especially Cabernet Sauvignon (6).

St Helena exhibits a wide diurnal temperature variability, reflecting the flat, valley-floor location that typifies many of its (especially older) vineyards. The climate estimate, with no site adjustments, is for these and their water-retentive alluvial soils. Sloping valley-edge soils are either gravelly or shallow over volcanic rock (see Halliday 1993; Swinchatt and Howell 2004) and their climates must be effectively more equable. Also the climates are cooler and more equable southwards towards Napa as marine influences ex San Francisco Bay and the Petaluma Gap become more prominent.

Walla Walla, Washington State, USA (Table A2.10)
Vineyard site adjustments, °C

Min.	Max.	
–0.2	–0.2	Average vineyard altitude c. 320 m
+0.4	–	Moderate slopes
+0.2	–	Some soils stony
+0.4	–0.2	Mean +0.1, range –0.6

Notes
Sun hours and cloud are from WMO (1998), R.H. as for Pendleton, Oregon, ex NCDC (1996).

Walla Walla's relative equability and high rainfall for its inland location result from maritime influences extending up the Columbia Valley, together with the presence of backing mountains. Other locations further upstream and in the Yakima Valley are cut off from these and are more arid.

Geneva, New York State, USA (Table A2.11)
Vineyard site adjustments, °C

Min.	Max.	
+0.2	+0.2	Most vineyards farther S
+0.1	+0.1	Average vineyard altitude c. 200 m
+0.4	–0.4	Most vines closer to lakes
+0.8	–	Steep slopes, E and W aspects
+1.5	–0.1	Mean +0.7, range –1.6

Notes
V_{min}, V_{max} and R.H. as for Syracuse, NY, sun hours and cloud as for Rochester, NY, all from WMO (1998). Main *vinifera* grape varieties are Chardonnay and Gewürztraminer (Maturity Group 3) and most famously Riesling (4).

All vineyards of necessity more or less directly overlook one or other of the Finger Lakes, to escape winter killing and spring frosts. See Chapter 4 for related discussion.

Santiago, Chile (Table A2.12)
Vineyard site adjustments: none.

Notes
Sun hours and cloud from Schwerdtfeger (1976) and Müller (1982). Leading grape varieties are Merlot (Maturity Group 5), Cabernet Sauvignon (6) and Carmenère (7).

Santiago Airport should fairly represent the valley-floor sites of the region's traditionally irrigated vineyards. These have a notably wide diurnal temperature range, tempered by restricted variability of the maxima (V_{max}) which results in very moderate high temperature extremes. Sites extending up adjacent hillsides with drip irrigation would have significantly higher minima and less frost risk.

Mendoza, Argentina (Table A2.13)
Vineyard site adjustments, °C

Min.	Max.	
−0.6	−0.6	Average vineyard altitude c. 800 m
−0.6	−0.6	Mean −0.6, range unchanged

Notes
Sun hours, cloud and rainfall ex WMO (1998); leading grape varieties
Malbec (Maturity Group 4), Cabernet Sauvignon (6) and Criolla and
Bonarda (= Croatina?) (both probably 7).

Based on mean temperatures, Mendoza is close to the warm limit for
superior table wines. But compensating for this is relative equability during
ripening, with warm nights and, for its temperature means, very moderate
extreme maxima, HVI and TVI, together with some cloudiness and
humidity. The combination allows production of good, full-bodied red
wines, but also indicates scope for expanding to cooler climates at locally
higher altitudes.

Blenheim, New Zealand (Table A2.14)
Vineyard site adjustments, °C

Min.	Max.	
−0.2	−0.2	Average vineyard altitudes c. 40 m
−	−0.4	Up-valley sea breezes
+0.4	−	Variably stony soils
+0.2	−0.6	Mean −0.2, range −0.8

Notes
Principal grape varieties are Sauvignon Blanc, Pinot Noir (both Maturity
Group 3) and Riesling (4). These make archetypal cool-climate wine styles,
as discussed in the introduction to this Appendix. Spring and autumn
frosts are a viticultural hazard.

Mildura, Victoria, Australia (Table A2.15)
Vineyard site adjustments, °C

Min.	Max.	
−	+0.4	Exposure to hot land winds
−	+0.4	Mean +0.2, range +0.4

Notes
Mildura is too far inland for cool sea breezes but is exposed to periodic
summer hot winds from central Australia that abnormally prolong afternoon heat. Main quality grape varieties are Chardonnay (Maturity Group

3), Shiraz (5) and Cabernet Sauvignon (6); a range of other varieties contribute to bulk wines.

Nuriootpa, South Australia (Table A2.16)

Vineyard site adjustments, °C

Min. Max.

+0.4	–	Undulating, vs valley floor
+0.4	–	Mean +0.2, range –0.4

Notes

Principal grape varieties are Chardonnay (Maturity Group 3), Semillon and Riesling (4), Shiraz and Merlot (5), Cabernet Sauvignon (6), Grenache (7) and Mourvèdre (8).

Nuriootpa is at the northern, upper end of the Barossa Valley. Declining altitudes southwards should more or less offset increasing latitude, giving comparable temperatures for the valley's lower elevations throughout. The Barossa is especially noted for its full-bodied red wines from Shiraz.

Launceston, Tasmania, Australia (Table A2.17)

Vineyard site adjustments, °C

Min.Max.

+0.1	+0.1	Average vineyard altitude c. 50 m
+0.6	–	Strongly more sloping
+0.2	–	ENE aspects
+0.8	–0.8	Closer to estuary, ocean
+1.7	–0.7	Mean +0.5, range –2.4

Notes

The raw data other than for rainfall are averages of those for 1965–89 from Launceston Mount Pleasant (41°28'S, 147°09'E, alt. 137 m), and for 1980–99 for Launceston Ti Tree Bend (41°25'S, 147°07'E, alt. 5 m) which is on flat valley floor. Rainfall records are for Ti Tree Bend only. All data are from BOMA (2000).

The adjustments are for the main vineyard sites downstream on steep slopes overlooking the Derwent Estuary, and/or close to the ocean (see Halliday 2006). The 4% upward adjustment to 3p.m. R.H. recognizes this. I make no adjustment for cooling by sea breezes, because marine winds come from the north and may equally be warming (see introduction to this Appendix).

Main grape varieties are Chardonnay, Pinot Noir, Pinot Gris, Sauvignon Blanc and Gewürztraminer (all Maturity Group 3) and Riesling (4). Cabernet Sauvignon (6) succeeds in a few of the warmest exposures.

Stellenbosch, South Africa (Table A2.18)

Vineyard site adjustments: none

Notes

Sun hours are as for Capetown D.F. Malan Airport, and so are possibly conservative. Otherwise the data probably represent fairly the older-established and more inland Stellenbosch vineyards. There is some influence of cool marine winds, but I assume these are largely cut off by mountains to the south-east.

TABLE A2.1

DIJON (Longvic Airport), **FRANCE**. 47°16'N, 5°05'E, altitude 227 m. Temperatures 1961–90 (WMO 1998)

Period	Raw temperatures, °C						Raw temperature indices			°days with 19° cut-off			Averages of			
	Lowest recorded minimum	Average lowest minimum	Average minimum	Average mean	Average maximum	Average highest maximum	Spring frost index	Heat stress index	Temp. variability index	Raw	Adj. for lat. & diurnal range	+ vine sites adj.*	Daily sun hours	Cloud cover, eighths	Total rain, mm	% R.H. 1.30 p.m.
April	−4.0	−1.1	5.0	9.9	14.7	21.8	11.0	11.9	32.6	12	23	33	5.9	5.4	52	58
May	−1.0	2.8	8.7	13.7	18.7	25.7	10.9	12.0	32.9	115	137	153	6.6	5.5	86	58
June	2.9	6.3	12.0	17.2	22.4	29.4		12.2	33.5	216	242	258	7.8	5.0	62	58
July	5.2	8.4	14.1	19.7	25.3	32.6		12.9	35.4	279	279	279	8.6	4.5	51	57
August	5.0	7.4	13.7	19.1	24.5	31.6		12.5	35.0	279	279	279	7.4	4.6	65	53
September	2.2	4.4	10.9	16.1	21.3	28.2		12.1	34.2	183	185	194	6.5	4.7	67	59
October	−4.7	−0.2	7.2	11.3	15.5	22.9		11.6	31.4	44	44	51	3.9	5.7	58	70
Apr.–Oct.		4.0	10.2	15.3	20.3	27.5		12.2	33.6	1128	1189	1247	6.7	5.1	441	59
Nov.–Mar.		−5.6	0.7	3.9	7.1	14.8			26.8						291	

Estimates for final 30 days to maturity

Maturity group	Maturity date	Lowest recorded minimum	Average lowest minimum	Average minimum	Average mean	Average maximum	Average highest maximum	Spring frost index	Heat stress index	Temp. variability index	Raw	Adj. for lat. & diurnal range	+ vine sites adj.*	Daily sun hours	Cloud cover, eighths	Total rain, mm	% R.H. 1.30 p.m.
1	3 September		*	14.1	19.1	24.1	31.2	*	12.1	33.3				7.3	4.6	65	53
2	11 September		*	13.4	18.4	23.3	30.4	*	12.0	33.1				6.9	4.6	66	53
3	20 September		*	12.6	17.5	22.4	29.4	*	11.9	32.9				6.6	4.7	66	55
4	1 October		*	11.4	16.2	21.0	27.9	*	11.7	32.5				6.2	4.7	67	59
5	–		–	–	–	–	–	–	–	–				–	–	–	–
6	–		–	–	–	–	–	–	–	–				–	–	–	–
7	–		–	–	–	–	–	–	–	–				–	–	–	–
8	–		–	–	–	–	–	–	–	–				–	–	–	–
9	–		–	–	–	–	–	–	–	–				–	–	–	–

* Adjusted for vineyard sites: minima +0.7°, maxima –0.1°, means +0.3°, SFI and HSI –0.4, TVI –1.6.

TABLE A2.2

BORDEAUX (Mérignac Airport), **FRANCE**. 44°50′N, 0°42′W, altitude 61 m. Temperatures 1961–90 (WMO 1998)

Period	Raw temperatures, °C						Raw temperature indices			°days with 19° cut-off			Averages of			
	Lowest recorded minimum	Average lowest minimum	Average minimum	Average mean	Average maximum	Average highest maximum	Spring frost index	Heat stress index	Temp. variability index	Raw	Adj. for lat. & diurnal range	+ vine sites adj.*	Daily sun hours	Cloud cover, eighths	Total rain, mm	% R.H. 4 p.m.
April	−3.0	0.6	6.3	11.3	16.3	23.6	10.7	12.3	33.0	39	55	75	6.3	4.7	72	58
May	−0.5	3.8	9.5	14.6	19.7	27.7	10.8	13.1	34.1	143	162	182	6.8	4.8	77	58
June	4.0	6.9	12.4	17.8	23.2	31.2		13.4	35.1	234	252	273	8.1	4.8	56	58
July	6.9	9.7	14.4	20.2	26.0	34.2		14.0	36.1	279	279	279	8.9	4.5	47	55
August	6.0	8.9	14.2	19.9	25.6	32.7		12.8	35.2	279	279	279	8.0	4.4	54	54
September	2.2	6.4	12.2	18.0	23.7	30.8		12.8	35.9	240	241	254	6.9	4.6	74	57
October	−1.7	2.5	9.1	14.0	18.9	25.8		11.8	33.1	124	122	134	5.3	4.8	88	68
Apr.–Oct.		5.5	11.2	16.5	21.9	29.4		12.9	34.6	1338	1390	1476	7.2	4.7	468	58
Nov.–Mar.		−3.6	3.5	7.5	11.5	18.3			29.9						455	

Estimates for final 30 days to maturity

Maturity group	Maturity date*	Average lowest minimum*	Average minimum*	Average mean*	Average maximum*	Average highest maximum*	Spring frost index*	Heat stress index*	Temp. variability index*	Raw	Adj.	+ vine	Daily sun hours	Cloud cover, eighths	Total rain, mm	% R.H. 4 p.m.
1	23 August	15.4	20.6	25.8	33.0			12.6	33.4				8.2	4.4	52	54
2	30 August	15.1	20.3	25.5	32.6			12.4	33.3				8.0	4.4	54	54
3	6 September	14.8	20.0	25.2	32.2			12.3	33.5				7.8	4.4	57	54
4	13 September	14.4	19.6	24.8	31.9			12.3	33.6				7.6	4.5	60	54
5	20 September	13.9	19.2	24.4	31.4			12.3	33.6				7.3	4.5	66	55
6	28 September	13.3	18.5	23.6	30.8			12.2	33.7				7.0	4.6	73	56
7	7 October	12.5	17.7	22.8	30.0			12.0	33.3				6.6	4.6	78	58
8	19 October	11.2	16.1	21.0	28.2			11.6	32.3				6.0	4.7	83	63
9	–	–	–	–	–			–	–				–	–	–	–

* Adjusted for vineyard sites: minima +0.9°, maxima −0.1°, means +0.4°, SFI and HSI −0.5, TVI −2.0.

TABLE A2.3

MARSEILLE (Marignane Airport), **FRANCE**. 43°27'N, 5°14'E, altitude 36 m. Temperatures 1961–90 (WMO 1998)

Period	Raw temperatures, °C						Raw temperature indices			°days with 19° cut-off			Averages of			
	Lowest recorded minimum	Average lowest minimum	Average minimum	Average mean	Average maximum	Average highest maximum	Spring frost index	Heat stress index	Temp. variability index	Raw	Adj. for lat. & diurnal range	+ vine sites adj.*	Daily sun hours	Cloud cover, eighths	Total rain, mm	% R.H. 4 p.m.
April	0.3	3.0	8.6	13.2	17.8	23.3	10.2	10.1	29.5	96	118	124	8.2	3.8	48	52
May	2.2	6.3	12.2	17.1	21.9	27.5	10.8	10.4	30.9	220	243	250	9.4	3.8	42	53
June	6.8	11.4	15.9	20.8	25.8	30.8		10.0	29.3	270	270	270	10.9	3.5	28	50
July	11.8	14.3	18.5	23.8	29.0	33.9		10.1	30.1	279	279	279	11.8	2.2	14	46
August	9.4	13.1	18.0	23.2	28.4	33.3		10.1	30.6	279	279	279	10.6	2.9	29	48
September	6.6	10.2	15.4	20.3	25.2	30.1		9.8	29.7	270	270	270	8.5	3.4	47	54
October	0.4	5.5	11.6	16.0	20.5	26.3		10.3	29.7	186	184	184	6.6	4.1	78	61
Apr.–Oct.		9.1	14.3	19.2	24.1	29.3		10.1	30.0	1600	1643	1656	9.4	3.4	286	52
Nov.–Mar.		−2.1	4.5	8.6	12.7	18.0			28.3						259	

Estimates for final 30 days to maturity

Maturity group	Maturity date	*		*	*	*	*	*				Daily sun hours	Cloud cover, eighths	Total rain, mm	% R.H. 4 p.m.
1	11 August	18.7	23.5	28.2	33.2		9.7	28.7				11.3	2.5	19	46
2	18 August	18.7	23.5	28.2	33.2		9.7	29.0				11.2	2.6	22	47
3	25 August	18.6	23.4	28.1	33.1		9.7	29.1				11.0	2.8	25	47
4	31 August	18.4	23.2	28.0	32.9		9.7	29.0				10.6	2.9	29	47
5	7 September	18.0	22.8	27.7	32.4		9.6	28.8				10.2	3.0	33	48
6	14 September	17.6	22.3	26.9	31.9		9.6	28.6				9.8	3.1	37	49
7	20 September	17.0	21.6	26.2	31.1		9.5	28.4				9.3	3.2	41	51
8	27 September	16.3	20.8	25.3	30.2		9.4	28.1				8.8	3.3	45	53
9	4 October	15.4	19.8	24.3	29.2		9.4	28.0				8.3	3.5	49	55

* Adjusted for vineyard sites: minima +0.4°, maxima −0.4°, means unchanged, SFI and HSI −0.4, TVI −1.6.

TABLE A2.4

GEISENHEIM (Rhine Valley), GERMANY. 49°59'N, 7°57'E, altitude 120 m.   Temperatures 1961–90 (WMO 1998)

Period	Raw temperatures, °C — Lowest recorded minimum	Average lowest minimum	Average minimum	Average mean	Average maximum	Average highest maximum	Raw temperature indices — Spring frost index	Heat stress index	Temp. variability index	°days with 19° cut-off — Raw	Adj. for lat. & diurnal range	+ vine sites adj.*	Averages of — Daily sun hours	Cloud cover, eighths	Total rain, mm	% R.H. 4 p.m.
April	−5.2	−1.3	4.8	9.5	14.2	23.1	10.8	13.6	33.8	9	20	30	5.5		39	53
May	−0.2	1.6	8.7	13.8	18.9	27.1	12.2	13.3	35.7	118	141	164	6.6		51	52
June	3.4	5.4	11.9	17.0	22.0	30.6		13.6	35.3	210	244	267	6.7		59	52
July	5.4	7.3	13.4	18.6	23.9	32.0		13.4	35.2	267	279	279	7.2		57	52
August	5.8	7.1	13.2	18.4	23.6	31.2		12.8	34.5	260	274	279	6.7		53	52
September	2.7	3.8	10.3	15.2	20.1	26.8		11.6	32.8	156	158	170	5.2		41	59
October	−2.5	0.0	6.6	10.4	14.3	21.5		11.1	29.2	25	24	35	3.2		42	70
Apr.–Oct.			9.8	14.7	19.6	27.5	12.8	12.8	33.8	1045	1140	1224	5.9		342	56
Nov.–Mar.		−7.0	0.5	3.3	6.1	12.9			25.5						207	

Estimates for final 30 days to maturity

Maturity group	Maturity date	Lowest recorded minimum	Average lowest minimum	Average minimum	Average mean	Average maximum	Average highest maximum	Spring frost index	Heat stress index	Temp. variability index	Raw	Adj. for lat. & diurnal range	+ vine sites adj.*	Daily sun hours	Cloud cover, eighths	Total rain, mm	% R.H. 4 p.m.
1	1 September	*	*	14.1	18.7	23.3	31.0	*	12.2	32.0				6.7		53	52
2	10 September	*	*	13.4	18.0	22.5	29.7	*	11.7	31.5				6.3		49	53
3	21 September	*	*	12.4	16.8	21.2	28.2	*	11.3	30.9				5.8		44	55
4	4 October	*	*	10.9	15.2	19.4	26.1	*	10.9	30.1				5.0		41	60
5	–	–	–	–	–	–	–	–	–	–	–	–	–	–	–	–	–
6	–	–	–	–	–	–	–	–	–	–	–	–	–	–	–	–	–
7	–	–	–	–	–	–	–	–	–	–	–	–	–	–	–	–	–
8	–	–	–	–	–	–	–	–	–	–	–	–	–	–	–	–	–
9	–	–	–	–	–	–	–	–	–	–	–	–	–	–	–	–	–

* Adjusted for vineyard sites: minima +1.0°, maxima −0.2°, means +0.4°, SFI and HSI −0.6, TVI −2.4.

TABLE A2.5

VIENNA (Airport), AUSTRIA. 48°07'N,16°34'E, altitude 190 m. Temperatures 1973–93 (NCDC 1996)

Period	Raw temperatures, °C — Lowest recorded minimum	Average lowest minimum	Average minimum	Average mean	Average maximum	Average highest maximum	Raw temperature indices — Spring frost index	Heat stress index	Temp. variability index	°days with 19° cut-off — Raw	Adj. for lat. & diurnal range	+ vine sites adj.*	Averages of — Daily sun hours	Cloud cover, eighths	Total rain, mm	% R.H. 4 p.m.
April	−3.9	−1.1	4.9	9.5	14.1	22.5	10.6	13.0	32.8	10	25	25	5.7	5.0	64	50
May	0.0	4.1	9.7	14.6	19.4	26.1	10.5	11.5	31.7	143	169	169	7.1	4.8	66	51
June	3.9	7.2	12.9	17.6	22.3	29.4		11.8	31.6	228	'265	265	7.4	4.8	66	52
July	6.7	9.8	14.8	19.8	24.7	31.8		12.0	31.9	279	279	279	7.9	4.2	64	48
August	5.0	9.1	14.8	19.8	24.8	31.6		11.8	32.5	279	279	279	7.4	4.2	68	48
September	2.8	5.8	11.5	16.0	20.6	28.0		12.0	31.3	180	182	176	5.7	4.4	74	54
October	−5.6	−0.9	6.3	10.2	14.1	22.6		12.4	31.3	27	27	23	4.4	4.6	78	62
Apr.–Oct.		4.9	10.7	15.4	20.0	27.4		12.1	31.9	1146	1226	1216	6.5	4.6	480	52
Nov.–Mar.		−8.3	−0.7	2.2	5.2	13.8			28.0						214	

Estimates for final 30 days to maturity

Maturity group	Maturity date	* (Avg minimum)	* (Avg mean)	* (Avg maximum)	* (Avg highest maximum)	* (Heat stress index)	* (Temp. variability index)	Daily sun hours	Cloud cover, eighths	Total rain, mm	% R.H. 4 p.m.
1	1 September	15.0	19.6	24.1	30.9	11.4	30.8	7.4	4.2	68	48
2	9 September	14.2	18.8	23.4	30.1	11.4	30.4	6.9	4.3	70	50
3	19 September	13.0	17.6	22.1	29.0	11.5	30.0	6.4	4.3	72	52
4	3 October	11.4	15.5	19.6	27.1	11.6	29.7	5.6	4.4	74	55
5	–	–	–	–	–	–	–	–	–	–	–
6	–	–	–	–	–	–	–	–	–	–	–
7	–	–	–	–	–	–	–	–	–	–	–
8	–	–	–	–	–	–	–	–	–	–	–
9	–	–	–	–	–	–	–	–	–	–	–

* Adjusted for vineyard sites: minima +0.2°, maxima −0.6°, means −0.2°, SFI and HSI −0.4, TVI −1.6.

TABLE A2.6

TORINO (Casella Airport), **ITALY**. 45°13′N, 7°39′E, altitude 287 m. Temperatures 1973–90 (NCDC 1996)

Period	Raw temperatures, °C						Raw temperature indices			°days with 19° cut-off			Averages of			
	Lowest recorded minimum	Average lowest minimum	Average minimum	Average mean	Average maximum	Average highest maximum	Spring frost index	Heat stress index	Temp. variability index	Raw	Adj. for lat. & diurnal range	+ vine sites adj.*	Daily sun hours	Cloud cover, eighths	Total rain, mm	% R.H. 4 p.m.
April	-2.2	0.3	6.2	11.1	16.0	22.9	10.8	11.8	32.4	36	44	65	6.0		104	56
May	1.1	5.1	10.8	15.6	20.3	26.3	10.5	10.7	30.7	174	200	216	6.3		120	61
June	6.1	8.8	14.5	19.6	24.7	30.1		10.5	31.5	270	270	270	7.3		98	57
July	6.1	11.6	17.2	22.4	27.7	32.2		9.8	31.1	279	279	279	8.4		67	55
August	7.8	10.9	16.7	21.6	26.6	30.9		9.3	29.9	279	279	279	7.2		80	58
September	2.8	8.1	13.5	18.4	23.3	28.2		9.8	29.9	252	254	263	5.6		70	60
October	-2.2	2.0	8.1	12.6	17.2	23.8		11.2	30.9	81	79	88	4.6		89	65
Apr.–Oct.		6.7	12.4	17.3	22.3	27.8		10.4	30.9	1371	1405	1460	6.5		628	59
Nov.–Mar.		-5.3	0.1	4.6	9.0	16.3			30.5						287	

Estimates for final 30 days to maturity

Maturity group	Maturity date															
1	22 August	*	17.7	22.4	27.1	31.4	*	9.0	28.7				7.6		76	57
2	28 August	*	17.5	22.1	26.7	31.0	*	8.9	28.4				7.3		79	58
3	4 September	*	17.2	21.6	26.1	30.5	*	8.9	28.3				7.0		80	58
4	11 September	*	16.7	21.0	25.4	30.0	*	9.0	28.2				6.7		78	59
5	17 September	*	16.0	20.3	24.7	29.4	*	9.1	28.2				6.3		75	59
6	24 September	*	15.1	19.5	23.9	28.8	*	9.3	28.3				5.9		72	60
7	3 October	*	13.9	18.3	22.7	27.8	*	9.5	28.4				5.5		71	60
8	15 October	*	12.0	16.5	21.0	26.4	*	9.9	28.7				5.1		78	62
9	—	—	—	—	—	—	—	—	—				—		—	—

* Adjusted for vineyard sites: minima +0.7°, maxima –0.1°, means +0.3°, SFI and HSI –0.4, TVI –1.6.

TABLE A2.7

JEREZ DE LA FRONTERA (Airport), SPAIN. 36°45'N, 6°04'W, altitude 30 m. Temperatures 1973–93 (NCDC 1996)

Period	Raw temperatures, °C						Raw temperature indices			°days with 19° cut-off			Daily sun hours	Averages of		
	Lowest recorded minimum	Average lowest minimum	Average minimum	Average mean	Average maximum	Average highest maximum	Spring frost index	Heat stress index	Temp. variability index	Raw	Adj. for lat. & diurnal range	+ vine sites adj.*		Cloud cover, eighths	Total rain, mm	% R.H. 4 p.m.
April	2.2	4.9	9.8	15.3	20.7	27.7	10.4	12.4	33.7	159	166	176	7.8		54	53
May	6.1	7.9	12.4	18.2	23.9	30.6	10.3	12.4	34.2	254	254	264	9.5		38	50
June	8.3	11.2	15.8	21.9	28.0	34.9		13.0	35.9	270	270	270	10.5		19	46
July	12.2	14.2	18.6	25.5	32.4	39.1		13.6	38.7	279	279	279	11.5		2	39
August	12.8	14.9	18.9	25.8	32.6	39.5		13.7	38.3	279	279	279	10.9		5	40
September	7.8	12.2	17.6	23.9	30.1	36.3		12.4	36.6	270	270	270	8.6		20	44
October	0.0	7.3	13.6	18.9	24.2	31.3		12.4	34.6	276	278	275	7.3		61	53
Apr.–Oct.		10.4	15.2	21.3	27.4	34.2		12.8	36.0	1787	1796	1813	9.4		199	46
Nov.–Mar.		1.5	7.8	12.7	17.6	22.8			31.1						447	

Estimates for final 30 days to maturity

Maturity group	Maturity date	*	*	*	*	*	*	*							
1	4 August	19.7	25.7	31.6	38.5		12.8	35.1				11.5		2	39
2	11 August	20.0	26.0	31.9	38.9		12.9	35.1				11.4		2	39
3	17 August	20.0	26.0	31.9	38.9		12.9	35.0				11.3		3	39
4	24 August	19.9	25.9	31.8	38.8		12.9	34.9				11.1		4	39
5	31 August	19.7	25.7	31.6	38.5		12.8	34.7				10.9		5	40
6	6 September	19.5	25.4	31.3	38.0		12.6	34.5				10.6		7	40
7	13 September	19.2	25.0	30.8	37.4		12.4	34.2				10.2		10	41
8	20 September	19.0	24.6	30.2	36.7		12.1	33.8				9.7		13	42
9	26 September	18.6	24.1	29.6	35.8		11.7	33.3				9.0		17	43

* Adjusted for vineyard sites: minima +0.8°, maxima –1.0°, means –0.1°, SFI and HSI –0.9, TVI –3.6..

TABLE A2.8

FRESNO (Airfield), CALIFORNIA. 36°13'N, 119°43'W, altitude 100 m. Temperatures 1949–95 (NCDC 1996)

Period	Raw temperatures, °C						Raw temperature indices			°days with 19° cut-off			Averages of			
	Lowest recorded minimum	Average lowest minimum	Average minimum	Average mean	Average maximum	Average highest maximum	Spring frost index	Heat stress index	Temp. variability index	Raw	Adj. for lat. & diurnal range	+ vine sites adj.*	Daily sun hours	Cloud cover, eighths	Total rain, mm	% R.H. 4 p.m.
April	0.0	3.6	8.6	16.1	23.7	32.1	12.5	16.0	43.6	183	158		11.2	3.5	26	35
May	2.2	6.5	12.0	20.3	28.5	36.9	13.8	16.6	46.9	279	279		12.6	2.4	9	27
June	6.7	10.3	15.4	24.3	33.1	40.4		16.1	47.8	270	270		13.5	1.6	3	23
July	10.0	13.5	18.3	27.5	36.7	41.6		14.1	46.5	279	279		13.7	1.2	0	22
August	9.4	13.1	17.4	26.5	35.6	40.7		14.2	45.8	279	279		12.7	1.5	1	24
September	2.8	10.0	14.9	23.7	32.4	38.8		15.1	46.3	270	270		11.3	1.8	5	27
October	–2.8	4.7	10.3	18.5	26.7	34.3		15.8	46.0	264	266		9.7	2.4	12	34
Apr.–Oct.		8.8	13.8	22.4	31.0	37.8		15.4	46.1	1824	1801		12.1	2.1	56	27
Nov.–Mar.		–1.0	4.4	10.1	15.8	23.1			35.5						220	

Estimates for final 30 days to maturity

Maturity group	Maturity date	Average minimum	Average mean	Average maximum	Average highest maximum	Heat stress index	Temp. variability index	Daily sun hours	Cloud cover, eighths	Total rain, mm	% R.H. 4 p.m.
1	4 August	18.3	27.6	36.8	41.7	14.1	46.4	13.7	1.2	0	22
2	11 August	18.2	27.4	36.6	41.6	14.1	46.3	13.5	1.3	0	23
3	18 August	18.0	27.2	36.3	41.4	14.1	46.2	13.3	1.3	1	23
4	24 August	17.7	26.9	36.0	41.1	14.2	46.0	13.0	1.4	1	24
5	31 August	17.4	26.5	35.6	40.7	14.2	45.8	12.7	1.5	1	24
6	7 September	16.9	26.0	35.1	40.3	14.3	45.9	12.4	1.6	1	25
7	13 September	16.4	25.4	34.5	39.9	14.5	46.0	12.1	1.6	2	25
8	20 September	15.8	24.8	33.8	39.5	14.7	46.1	11.8	1.7	3	26
9	27 September	15.2	24.1	33.0	39.1	15.0	46.2	11.5	1.8	5	27

* No adjustments for vineyard sites: airfield representative of central San Joaquin Valley.

TABLE A2.9

SAINT HELENA (Napa Valley), **CALIFORNIA**. 38°30'N, 122°28'W, altitude 70 m.   Temperatures 1931–2004 (WRCC 2005)

Period	Raw temperatures, °C — Lowest recorded minimum	Average lowest minimum	Average minimum	Average mean	Average maximum	Average highest maximum	Raw temperature indices — Spring frost index	Heat stress index	Temp. variability index	°days with 19° cut-off — Raw	Adj. for lat. & diurnal range	+ vine sites adj.*	Averages of — Daily sun hours	Cloud cover, eighths	Total rain, mm	% R.H.
April	−2.8	1.2	5.9	13.9	21.9	29.4	12.7	15.5	44.2	117	87		10.0	4.2	53	
May	−1.1	3.4	8.4	17.0	25.7	34.1	13.6	17.1	48.0	217	175		11.8	3.1	19	
June	2.2	6.1	10.5	19.8	29.2	37.8		18.0	50.4	270	242		13.1	2.4	6	
July	3.9	7.9	11.4	21.7	32.1	38.5		16.8	51.3	279	279		13.3	1.7	1	
August	0.0	7.8	11.2	21.4	31.7	37.9		16.5	50.6	279	279		12.3	2.2	2	
September	1.7	5.5	9.8	20.0	30.2	37.1		17.1	52.0	270	270		10.8	2.4	7	
October	−5.0	2.4	7.7	16.5	25.3	32.8		16.3	48.0	202	202		8.9	3.0	47	
Apr.–Oct.		4.9	9.3	18.6	28.0	35.4		16.8	49.2	1634	1534		11.5	2.7	135	
Nov.–Mar.		−1.5	3.7	10.0	16.3	22.7			36.8						753	

Estimates for final 30 days to maturity

Maturity group	Maturity date	Average minimum	Average mean	Average maximum	Average highest maximum	Heat stress index	Temp. variability index	Daily sun hours	Cloud cover, eighths	Total rain, mm
1	27 August	11.3	21.5	31.8	38.0	16.5	50.7	12.5	2.1	2
2	2 September	11.1	21.3	31.6	37.9	16.6	50.6	12.2	2.2	2
3	9 September	10.9	21.1	31.4	37.8	16.7	50.8	11.9	2.2	3
4	16 September	10.6	20.8	31.1	37.6	16.8	51.2	11.6	2.3	4
5	22 September	10.3	20.5	30.7	37.4	16.9	51.6	11.3	2.3	5
6	29 September	9.9	20.1	30.3	37.1	17.0	51.9	10.9	2.4	7
7	6 October	9.4	19.6	29.8	36.5	16.9	51.5	10.4	2.5	10
8	14 October	8.9	18.9	28.9	35.5	16.6	50.6	9.9	2.7	15
9	26 October	8.1	17.5	26.8	33.9	16.4	49.0	9.2	2.9	35

* No temperature adjustments for valley-floor vineyard sites; for others, see Site Notes.

TABLE A2.10

WALLA WALLA (Airport), WASHINGTON STATE. 46°02'N, 118°20'W, altitude 289 m. Temperatures 1950–88 (NCDC 1996)

Period	Lowest recorded minimum	Average lowest minimum	Average minimum	Average mean	Average maximum	Average highest maximum	Spring frost index	Heat stress index	Temp. variability index	°days Raw	Adj. for lat. & diurnal range	+ vine sites adj.*	Daily sun hours	Cloud cover, eighths	Total rain, mm	% R.H. 4 p.m.
April	−2.8	0.7	5.9	11.4	17.0	26.3	10.7	14.9	36.7	43	50	58	7.4	5.4	34	42
May	−2.2	4.2	9.6	15.7	21.7	31.8	11.5	16.1	39.7	177	183	191	9.3	4.8	36	37
June	5.0	8.1	13.3	19.8	26.4	35.4		15.6	40.4	270	270	270	10.9	4.1	25	32
July	7.8	11.0	16.7	24.0	31.4	39.2		15.2	42.9	279	279	279	12.4	2.6	11	23
August	7.8	11.2	16.3	23.3	30.3	38.1		14.8	40.9	279	279	279	11.1	2.7	16	26
September	1.1	5.3	11.8	18.4	25.0	33.6		15.2	41.5	252	254	257	9.0	3.5	22	32
October	−5.6	0.5	6.9	12.3	17.8	26.4		14.1	36.8	71	71	73	6.4	4.5	38	46
Apr.–Oct.		5.9	11.5	17.8	24.2	33.0		15.1	39.8	1371	1386	1407	9.5	3.9	182	34
Nov.–Mar.		−7.4	0.8	4.4	8.1	17.8			32.5						241	

Estimates for final 30 days to maturity

Maturity group	Maturity date	Average minimum *	Average mean *	Average maximum *	Average highest maximum *	Spring frost index *	Heat stress index *	Temp. variability index *	Daily sun hours	Cloud cover, eighths	Total rain, mm	% R.H. 4 p.m.
1	25 August	16.8	23.5	30.3	38.1	*	14.6	40.0	11.3	2.5	15	26
2	1 September	16.5	23.2	30.0	37.8	*	14.5	39.7	11.1	2.7	16	26
3	7 September	15.9	22.5	29.0	37.2	*	14.6	39.7	10.7	2.9	17	27
4	14 September	15.1	21.5	27.8	36.3	*	14.7	39.8	10.3	3.1	18	28
5	22 September	13.9	20.2	26.4	35.1	*	14.8	40.1	9.7	3.3	20	30
6	30 September	12.3	18.6	24.8	33.5	*	14.9	40.3	9.0	3.5	22	32
7	11 October	10.4	16.3	22.2	30.8	*	14.5	39.6	8.2	3.9	26	36
8	—	—	—	—	—	—	—	—	—	—	—	—
9	—	—	—	—	—	—	—	—	—	—	—	—

* Adjusted for vineyard sites: minima +0.4°, maxima −0.2°, means +0.1°, SFI and HSI −0.3, TVI −1.2.

TABLE A2.11

GENEVA (Research Farm), NEW YORK STATE. 42°53'N, 77°02'W, altitude 219 m. Temperatures 1961–90 (WMO 1998)

Period	Raw temperatures, °C						Raw temperature indices			°days with 19° cut-off			Averages of			
	Lowest recorded minimum	Average lowest minimum	Average minimum	Average mean	Average maximum	Average highest maximum	Spring frost index	Heat stress index	Temp. variability index	Raw	Adj. for lat. & diurnal range	+ vine sites adj.*	Daily sun hours	Cloud cover, eighths	Total rain, mm	% R.H. 4 p.m.
April		-5.8	2.0	7.2	12.3	24.8	13.0	17.6	40.9	0	0	8	7.2	5.4	74	53
May		0.2	7.7	13.4	19.8	29.4	13.2	16.0	41.3	105	107	143	8.6	5.2	76	53
June		5.4	12.8	18.4	24.1	31.7		13.3	37.6	252	263	270	9.9	4.9	93	54
July		9.6	15.7	21.3	26.9	32.8		11.5	34.4	279	279	279	10.1	4.7	75	54
August		7.8	14.8	20.3	25.8	32.0		11.7	35.2	279	279	279	8.8	5.0	80	57
September		2.4	10.8	16.3	21.7	30.3		14.0	38.8	189	190	210	7.1	5.1	82	60
October		-3.2	4.9	10.1	15.2	25.3		15.2	38.8	21	21	33	5.0	5.6	74	60
Apr.–Oct.		2.3	9.8	15.3	20.8	29.5		14.2	38.1	1125	1139	1222	8.1	5.1	554	56
Nov.–Mar.		-17.4	-5.8	-1.5	2.8	15.0			41.0						283	

Estimates for final 30 days to maturity

Maturity group	Maturity date	Average lowest minimum	Average minimum	Average mean	Average maximum	Average highest maximum	Spring frost index	Heat stress index	Temp. variability index	Raw	Adj. for lat. & diurnal range	+ vine sites adj.*	Daily sun hours	Cloud cover, eighths	Total rain, mm	% R.H. 4 p.m.
		*	*	*	*	*	*	*	*							
1	5 September	15.9	20.5	25.0	31.7			11.3	33.1				8.5	5.0	81	58
2	13 September	15.0	19.5	24.1	31.3			11.9	34.3				8.1	5.0	82	58
3	22 September	13.7	18.3	22.9	30.7			12.6	35.2				7.6	5.1	82	59
4	3 October	11.8	16.5	21.2	29.8			13.4	35.7				6.9	5.7	82	60
5	–	–	–	–	–	–	–	–	–				–	–	–	–
6	–	–	–	–	–	–	–	–	–				–	–	–	–
7	–	–	–	–	–	–	–	–	–				–	–	–	–
8	–	–	–	–	–	–	–	–	–				–	–	–	–
9	–	–	–	–	–	–	–	–	–				–	–	–	–

* Adjusted for vineyard sites: minima +1.5°, maxima −0.1°, means +0.7°, SFI and HSI −0.8, TVI −3.2.

TABLE A2.12

SANTIAGO (Los Ocerrillos Airport), **CHILE**. 33°23'S, 70°47'W, altitude 474 m. Temperatures 1973–93 (NCDC 1996)

Period	Raw temperatures, °C						Raw temperature indices			°days with 19° cut-off			Averages of			
	Lowest recorded minimum	Average lowest minimum	Average minimum	Average mean	Average maximum	Average highest maximum	Spring frost index	Heat stress index	Temp. variability index	Raw	Adj. for lat. & diurnal range	+ vine sites adj.*	Daily sun hours	Cloud cover, eighths	Total rain, mm	% R.H. 2 p.m.
October	−2.8	1.4	7.2	14.6	21.9	29.4	13.2	14.8	42.7	143	119		6.9	3.6	13	49
November	0.0	4.1	9.1	17.0	25.0	31.9	12.9	14.9	43.7	210	175		8.8	2.8	5	42
December	0.0	6.2	10.8	19.3	27.8	33.3	13.1	14.0	44.1	279	239		10.5	1.7	5	38
January	6.1	8.3	12.2	20.8	29.3	33.8		13.0	42.6	279	279		10.7	1.5	0	37
February	5.0	8.1	11.4	20.1	28.8	33.1		13.0	42.4	252	252		9.9	1.5	3	39
March	0.0	5.2	9.7	18.1	26.6	32.0		13.9	43.7	251	250		8.7	1.7	5	43
April	−1.7	1.3	7.1	14.6	22.2	29.7		15.1	43.5	138	140		6.5	3.0	13	49
Oct.–Apr.		4.9	9.6	17.8	25.9	31.9		14.1	43.2	1552	1454		8.9	2.3	44	42
May–Sept.		−1.9	4.0	10.0	16.0	23.9			37.8						295	

Estimates for final 30 days to maturity

Maturity group	Maturity date			Average minimum	Average mean	Average maximum	Average highest maximum		Heat stress index	Temp. variability index				Daily sun hours	Cloud cover, eighths	Total rain, mm	% R.H. 2 p.m.
1	24 February			11.6	20.3	29.0	33.3		13.0	42.4				10.0	1.5	2	39
2	2 March			11.4	20.1	28.8	33.1		13.0	42.4				9.9	1.5	3	39
3	9 March			11.2	19.8	28.4	32.9		13.1	42.6				9.7	1.6	3	40
4	16 March			10.8	19.4	27.9	32.6		13.2	42.9				9.4	1.6	4	41
5	24 March			10.3	18.8	27.2	32.3		13.5	43.4				9.1	1.7	4	42
6	2 April			9.6	18.0	26.4	31.9		13.9	43.7				8.6	1.8	5	43
7	12 April			8.6	16.9	25.2	31.3		14.4	43.6				8.0	2.1	7	45
8	27 April			7.5	15.2	23.0	30.1		14.9	43.5				6.8	2.8	12	48
9	–			–	–	–	–		–	–				–	–	–	–

* No vineyard site adjustments: airport fairly represents traditional valley-floor sites.

TABLE A2.13

MENDOZA (El Plumeril Airport), **ARGENTINA**. 32°50'S, 68°47'W, altitude 704 m. Temperatures 1972–93 (NCDC 1996)

Period	Raw temperatures, °C — Lowest recorded minimum	Average lowest minimum	Average minimum	Average mean	Average maximum	Average highest maximum	Raw temperature indices — Spring frost index	Heat stress index	Temp. variability index	°days with 19° cut-off — Raw	Adj. for lat. & diurnal range	+ vine sites adj.*	Averages of — Daily sun hours	Cloud cover, eighths	Total rain, mm	% R.H. 3 p.m.
October	0.0	4.4	11.9	18.4	24.8	33.5	14.0	15.1	42.0	260	248	229	9.1	3.1	11	32
November	2.2	7.3	14.9	21.5	28.0	36.3	14.2	14.8	42.1	270	270	270	9.8	3.1	16	32
December	7.2	12.1	18.2	24.3	30.3	36.9		12.6	36.9	279	279	279	9.2	3.1	24	34
January	10.0	12.9	19.1	25.1	31.1	37.5		12.4	36.6	279	279	279	9.6	3.2	36	37
February	9.4	12.4	18.1	23.9	29.8	35.8		11.9	35.1	252	252	252	9.2	3.3	34	40
March	2.8	8.9	15.9	21.4	26.9	34.0		12.6	36.1	279	279	279	7.6	3.4	27	46
April	−1.7	3.6	11.1	16.9	22.7	29.2		12.3	37.2	207	211	194	7.3	3.2	13	46
Oct.–Apr.		8.8	15.6	21.6	27.7	34.7		13.1	38.0	1826	1818	1782	8.8	3.2	161	38
May–Sept.		−1.0	4.8	11.0	17.2	25.6			39.0						28	

Estimates for final 30 days to maturity

Maturity group	Maturity date	Lowest recorded minimum *	Average lowest minimum *	Average minimum *	Average mean *	Average maximum *	Average highest maximum *	Heat stress index	Temp. variability index	Daily sun hours	Cloud cover, eighths	Total rain, mm	% R.H. 3 p.m.
1	24 January			18.5	24.5	30.5	36.9	12.4	36.6	9.6	3.2	35	37
2	3 February			18.5	24.5	30.5	36.9	12.4	36.5	9.6	3.2	36	37
3	10 February			18.3	24.3	30.3	36.6	12.3	36.3	9.5	3.2	36	38
4	16 February			18.0	24.0	30.0	36.1	12.1	35.9	9.4	3.3	35	38
5	23 February			17.8	23.7	29.6	35.7	12.0	35.4	9.3	3.3	35	39
6	2 March			17.5	23.3	29.2	35.2	11.9	35.1	9.2	3.3	34	40
7	8 March			17.1	22.9	28.8	34.8	11.9	35.2	9.0	3.3	33	41
8	15 March			16.7	22.4	28.2	34.4	12.0	35.4	8.6	3.3	32	42
9	22 March			16.2	21.8	27.4	34.0	12.2	35.7	8.1	3.4	30	44

* Adjusted for vineyard sites: minima, maxima and means −0.6°; SFI, HSI and TVI unchanged.

TABLE A2.14

BLENHEIM (Airport), **NEW ZEALAND**. 41°31'S, 173°57'E, altitude 4 m. Temperatures 1971–2000 (NIWA 2003)

Period	Raw temperatures, °C						Raw temperature indices			°days with 19° cut-off			Averages of			
	Lowest recorded minimum	Average lowest minimum	Average minimum	Average mean	Average maximum	Average highest maximum	Spring frost index	Heat stress index	Temp. variability index	Raw	Adj. for lat. & diurnal range	+ vine sites adj.*	Daily sun hours	Cloud cover, eighths	Total rain, mm	% R.H. 3 p.m.
October	−2.1	−0.1	7.5	12.8	18.1	24.5	12.9	11.7	35.2	87	98	98	7.2		70	60
November	−1.7	1.5	9.6	14.9	20.1	26.7	13.4	11.8	35.7	147	160	160	7.8		43	56
December	0.9	4.0	11.4	16.7	22.0	28.9	12.7	12.2	35.5	208	221	221	8.2		54	53
January	1.7	4.8	12.5	18.2	23.8	31.8		13.6	38.3	254	262	262	8.3		47	53
February	0.0	5.0	12.2	17.9	23.6	31.0		13.1	37.4	221	223	217	8.1		27	55
March	−0.6	3.2	11.4	16.7	22.0	27.7		11.0	35.1	208	208	202	7.2		54	55
April	−2.6	1.0	8.4	13.7	19.0	24.6		10.9	34.2	111	111	105	6.2		64	59
Oct.–Apr.		2.8	10.4	15.8	21.2	27.9		12.0	35.9	1236	1283	1265	7.6		359	56
May–Sept.	−2.6	3.7	9.0	14.3	18.9				32.1						299	

Estimates for final 30 days to maturity

Maturity group	Maturity date	*	*	*	*		*	*								
1	9 March	12.4	17.6	22.8	29.7		12.1	35.1				7.9		37	55
2	18 March	12.2	17.3	22.3	28.6		11.3	34.3				7.6		45	55
3	28 March	11.8	16.7	21.7	27.5		10.8	33.6				7.3		54	55
4	9 April	10.7	15.6	20.5	26.2		10.6	33.2				6.9		60	56
5	28 April	8.9	13.8	18.7	24.3		10.5	32.7				6.3		64	59
6	–	–	–	–	–		–	–				–		–	–
7	–	–	–	–	–		–	–				–		–	–
8	–	–	–	–	–		–	–				–		–	–
9	–	–	–	–	–		–	–				–		–	–

* Adjusted for vineyard sites: minima +0.2°, maxima –0.6°, means –0.2°, SFI and HSI –0.4, TVI –1.6.

TABLE A2.15

MILDURA (Airport), **VICTORIA**. 34°14'S, 142°05'E, altitude 50 m. Temperatures 1957–2000 (BOMA 2000)

Period	Raw temperatures, °C						Raw temperature indices			°days with 19° cut-off			Averages of			
	Lowest recorded minimum	Average lowest minimum	Average minimum	Average mean	Average maximum	Average highest maximum	Spring frost index	Heat stress index	Temp. variability index	Raw	Adj. for lat. & diurnal range	+ vine sites adj.*	Daily sun hours	Cloud cover, eighths	Total rain, mm	% R.H. 3 p.m.
October	1.7	4.0	9.9	16.9	23.9	34.1	12.9	17.2	44.1	214	195	198	8.9	3.9	30	33
November	3.6	6.0	12.4	19.9	27.3	38.2	13.9	18.3	47.1	270	267	270	9.7	3.8	25	28
December	6.7	8.6	14.8	22.4	30.0	40.1		17.7	46.7	279	279	279	10.5	3.3	24	26
January	7.6	10.4	16.5	24.2	31.9	41.5		17.3	46.5	279	279	279	10.5	3.0	25	27
February	5.2	10.0	16.4	24.0	31.6	40.3		16.3	45.5	252	252	252	10.2	2.8	20	28
March	3.8	7.4	13.8	21.1	28.3	36.8		15.7	43.9	279	279	279	9.1	3.0	17	32
April	0.6	4.2	10.1	16.8	23.5	31.8		15.0	41.0	204	207	213	7.8	3.4	20	38
Oct.–Apr.		7.2	13.4	20.8	28.1	37.5		16.8	45.0	1777	1758	1770	9.5	3.3	161	30
May–Sept.		0.6	5.9	11.7	17.5	24.3			35.3						131	

Estimates for final 30 days to maturity

Maturity group	Maturity date	*	*	*	*		*	*								
1	30 January	16.5	24.4	32.3	41.9		17.5	47.3				10.5	3.0	24	27
2	6 February	16.7	24.6	32.6	41.9		17.3	47.2				10.5	3.0	24	27
3	13 February	16.8	24.7	32.6	41.7		17.0	47.0				10.4	2.9	23	27
4	19 February	16.7	24.6	32.4	41.4		16.8	46.8				10.4	2.9	22	28
5	26 February	16.5	24.3	32.2	40.9		16.6	46.5				10.3	2.8	22	28
6	5 March	16.1	23.9	31.7	40.3		16.4	46.1				10.1	2.8	21	29
7	11 March	15.7	23.4	31.2	39.6		16.2	45.8				9.9	2.9	19	29
8	18 March	15.1	22.8	30.5	38.9		16.1	45.4				9.6	2.9	18	30
9	25 March	14.4	22.0	29.5	38.0		16.0	45.0				9.3	3.0	17	31

* Adjusted for vineyard sites: minima unchanged, maxima +0.4°, means +0.2°, SFI and HSI +0.2, TVI +0.8.

TABLE A2.16

NURIOOTPA (Barossa Valley), **SOUTH AUSTRALIA**. 34°29'S, 139°00'E, altitude 274 m. Temperatures 1957–99 (BOMA 2000)

Period	Raw temperatures, °C						Raw temperature indices			°days with 19° cut-off			Averages of			
	Lowest recorded minimum	Average lowest minimum	Average minimum	Average mean	Average maximum	Average highest maximum	Spring frost index	Heat stress index	Temp. variability index	Raw	Adj. for lat. & diurnal range	+ vine sites adj.*	Daily sun hours	Cloud cover, eighths	Total rain, mm	% R.H. 3 p.m.
October	−0.6	1.8	8.0	14.2	20.4	30.8	12.4	16.6	41.4	130	125	134	7.9	4.5	48	45
November	−1.0	3.0	10.0	17.0	24.0	35.6	14.0	18.6	46.6	210	190	198	8.8	4.3	29	36
December	2.6	5.0	12.0	19.2	26.4	37.4		18.2	46.8	279	257	266	9.3	3.9	24	33
January	2.2	6.5	13.6	21.2	28.8	39.4		18.2	48.1	279	279	279	9.9	3.2	19	31
February	3.7	7.0	13.9	21.3	28.6	38.4		17.1	46.1	252	252	252	9.5	2.9	19	31
March	0.8	5.0	11.8	18.8	25.7	35.3		16.5	44.2	273	271	278	8.2	3.5	23	37
April	0.2	2.5	8.9	15.2	21.4	30.6		15.4	40.6	156	158	164`	6.6	4.2	37	44
Oct.–Apr.		4.4	11.2	18.1	25.0	35.3		17.2	44.8	1579	1532	1571	8.6	3.8	199	37
May– Sept.		−0.6	5.3	10.3	15.2	21.7		15.8	32.2						296	

Estimates for final 30 days to maturity

Maturity group	Maturity date	*	*			*	*
1	16 February	14.4	21.6	28.9	38.9	17.3	46.3
2	23 February	14.5	21.6	28.8	38.7	17.1	45.8
3	2 March	14.3	21.5	28.6	38.4	16.9	45.2
4	8 March	14.0	21.2	28.3	38.0	16.8	44.8
5	15 March	13.6	20.7	27.7	37.4	16.7	44.4
6	22 March	13.1	20.0	26.9	36.5	16.5	44.0
7	29 March	12.5	19.2	26.0	35.5	16.3	43.5
8	6 April	11.7	18.2	24.8	34.3	16.1	43.0
9	16 April	10.6	17.0	23.4	32.8	15.8	42.4

* Adjusted for vineyard sites: minima +0.4°, maxima unchanged, means +0.2°, SFI and HSI –0.2, TVI –0.8.

TABLE A2.17

LAUNCESTON, TASMANIA. 41°25'S, 147°07'E. Averages of Mount Pleasant and Ti Tree Bend sites: see Site Notes

Period	Raw temperatures, °C						Raw temperature indices			°days with 19° cut-off			Daily sun hours	Averages of		
	Lowest recorded minimum	Average lowest minimum	Average minimum	Average mean	Average maximum	Average highest maximum	Spring frost index	Heat stress index	Temp. variability index	Raw	Adj. for lat. & diurnal range	+ vine sites adj.*		Cloud cover, eighths	Total rain, mm	% R.H. 3 p.m.
October	−2.9	0.7	6.7	12.0	17.4	23.4	11.3	11.4	33.4	62	73	107	7.1	5.0	53	55
November	−3.0	2.2	8.4	14.0	19.7	26.3	11.8	12.3	35.4	120	127	160	7.6	5.0	51	53
December	1.2	4.2	10.3	16.0	21.7	28.8		12.8	36.0	186	193	227	8.1	4.9	49	49
January	2.1	5.9	11.6	17.6	23.7	30.6		13.0	36.8	236	237	272	8.6	4.6	47	49
February	1.6	5.1	11.7	17.9	24.1	30.8		12.9	38.1	221	223	237	8.4	4.5	31	47
March	−0.3	3.4	10.2	16.0	21.9	28.4		12.4	36.7	186	186	202	6.7	4.8	35	52
April	−4.1	0.8	7.6	12.9	18.2	23.6		10.7	33.4	87	87	102	5.7	4.8	55	58
Oct.–Apr.		3.2	9.5	15.2	21.0	27.4		12.2	35.7	1098	1126	1307	7.5	4.8	321	52
May–Sept.		−1.8	3.8	8.6	13.4	17.3			28.7						355	

Estimates for final 30 days to maturity

Maturity group	Maturity date	*	*	*	*	*	*	*					*
1	3 March	13.3	18.3	23.3	30.0	11.7	33.3				8.3	4.5	33	51
2	11 March	13.1	18.0	23.0	29.6	11.6	33.2				8.0	4.6	33	52
3	20 March	12.6	17.5	22.4	29.0	11.5	32.8				7.3	4.7	33	54
4	31 March	11.9	16.6	21.3	27.8	11.2	32.0				6.7	4.8	34	56
5	13 April	10.8	15.2	19.6	25.7	10.5	30.5				6.3	4.8	40	58
6	8 May	8.8	12.8	16.8	21.9	9.1	27.8				5.5	4.9	60	64
7														
8														
9														

* Adjusted for vineyard sites: minima +1.7°, maxima −0.7°, means +0.5°, SFI and HSI −1.2, TVI −4.8, R.H. +4%...

TABLE A2.18

STELLENBOSCH (Elsenburg Agricultural College), SOUTH AFRICA. 33°51'S, 18°50'E, altitude 162 m. Temperatures 1961–87 (SAWS 2003)

Period	Raw temperatures, °C						Raw temperature indices			°days with 19° cut-off			Averages of			
	Lowest recorded minimum	Average lowest minimum	Average minimum	Average mean	Average maximum	Average highest maximum	Spring frost index	Heat stress index	Temp. variability index	Raw	Adj. for lat. & diurnal range	+ vine sites adj.*	Daily sun hours	Cloud cover, eighths	Total rain, mm	% R.H. 2 p.m.
October	2.7	4.6	9.6	15.8	22.1	32.3	11.2	16.5	40.2	180	173		9.0	2.4	40	51
November	3.7	6.2	11.5	18.3	25.2	34.1	12.1	15.8	41.6	249	229		10.3	1.7	24	46
December	6.4	8.0	13.2	20.0	26.7	35.2		15.2	40.7	279	279		10.8	1.7	21	46
January	6.4	9.2	14.1	21.0	27.9	36.5		15.5	41.1	279	279		10.9	1.6	18	45
February	7.6	9.2	14.3	21.4	28.5	36.7		15.3	41.7	252	252		10.6	1.4	17	44
March	5.7	7.8	13.1	20.1	27.1	35.9		15.8	42.1	279	279		9.4	1.7	25	44
April	4.6	6.3	11.4	17.7	24.0	32.9		15.2	39.2	231	235		7.8	2.4	47	49
Oct.–Apr.		7.3	12.5	19.2	25.9	34.8		15.6	40.9	1749	1726		9.8	1.8	192	46
May–Sept.		2.9	7.8	12.9	18.1	26.5			33.9						403	

Estimates for final 30 days to maturity

Maturity group	Maturity date	Average minimum	Average mean	Average maximum	Average highest maximum	Heat stress index	Temp. variability index	Daily sun hours	Cloud cover, eighths	Total rain, mm	% R.H. 2 p.m.
1	7 February	14.3	21.3	28.3	36.8	15.5	41.3	10.9	1.5	18	45
2	14 February	14.5	21.5	28.5	36.9	15.4	41.4	10.9	1.5	17	45
3	20 February	14.5	21.5	28.6	36.9	15.4	41.5	10.8	1.4	17	44
4	27 February	14.4	21.4	28.5	36.7	15.3	41.7	10.7	1.4	17	44
5	6 March	14.2	21.2	28.3	36.6	15.4	41.8	10.5	1.4	17	44
6	12 March	14.0	21.0	28.1	36.5	15.5	41.9	10.3	1.4	18	44
7	19 March	13.7	20.7	27.7	36.3	15.6	42.0	10.0	1.5	20	44
8	26 March	13.4	20.4	27.4	36.1	15.7	42.0	9.7	1.6	22	44
9	1 April	13.0	20.0	26.9	35.8	15.8	42.1	9.3	1.7	25	44

* No vineyard site adjustments: Elsenburg data appear representative for the vineyards inland from Stellenbosch to Paarl.

References

Abarea-Del-Rio, R. and Mestre, O. (2006) Decadal to secular time scales variability in temperature measurements over France. Geophysical Research Letters 33, L13705, doi:10.1029/2006GL026019.

Acock, B., Acock, M.C. and Pasternak, D. (1990) Interactions of CO_2 enrichment and temperature on carbohydrate production and accumulation in muskmelon leaves. Journal of the American Society for Horticultural Science 115, 525–529.

Adams, L.D. (1984) The Wines of America, third edition, Sidgwick & Jackson, London.

Adams, S.R., Cockshull, K.E. and Cave, C.R.J. (2001) Effect of temperature on the growth and development of tomato fruits. Annals of Botany 88, 869–877.

Allan, W. (1998) The practical implications of organic viticulture. Australian Viticulture 2 (5), 49–59.

Allen, L.H. (1971) Variations in carbon dioxide concentration over an agricultural field. Agricultural Meteorology 8, 5–24.

Allen, L.H., Stewart, D.W. and Lemon, E.R. (1974) Photosynthesis in plant canopies: effect of light response curves and radiation source geometry. Photosynthetica 8, 184–207.

Alleweldt, G. and Düring, H. (1972) Influence de la photopériode sur la croissance et la teneur en acide abscissique. Vitis 11, 280–288.

Amerine, M.A. and Winkler, A.J. (1944) Composition and quality of musts and wines of California grapes. Hilgardia 15, 493–675.

Anderson, M., Fidelibus, M. and Waterhouse, A. (2008) Effects of abscisic acid on phenolic composition of Cabernet Sauvignon and Merlot winegrapes. American Journal of Enology and Viticulture 59, 337A. (Abstract from ASEV 59th Annual Meeting 2008.)

Anderson, T.L., Charlson, R.J. Schwartz, S.E., Knutti, R., Boucher, O., Rodhe, H. and Heintzenberg, J. (2003) Climate forcing by aerosols – a hazy picture. Science 300, 1103–1104.

Angus, N.S., O'Keeffe, T.J., Stuart, K.R. and Miskelly, G.M. (2006) Regional classification of New Zealand red wines using inductively-coupled plasma-mass spectrometry (ICP-MS). Australian Journal of Grape and Wine Research 12, 170–176.

Antolin, M.C., Baigorri, H., de Luis, I., Aguirrezabel, F., Geny, L., Broquedis, M. and Sanchez-Diaz, M. (2003) ABA during reproductive development in non-irrigated grapevines (*Vitis vinifera* L. cv. Tempranillo). Australian Journal of Grape and Wine Research 9, 169–176.

Antolin, M.C., Ayari, M. and Sanchez-Diaz, M. (2006) Effects of partial rootzone drying on yield, ripening and berry ABA in potted Tempranillo grapevines with split roots. Australian Journal of Grape and Wine Research 12, 13–20.

Antolin, M.C., Santesteban, H., Santa Maria, E., Aguirreolea, J. and Sanchez-Diaz, M. (2008) Involvement of abscisic acid and polyamines in berry ripening of *Vitis vinifera* (L.) subjected to water deficit irrigation. Australian Journal of Grape and Wine Research 14, 123–133.

Ashenfelter, O., Ashmore, D. and Lalonde, R. (1995) Bordeaux wine vintage quality and the weather. Chance (New York) 8 (4), 7–14.

Asselin, C., Morlat, R. and Salette, J. (1996) Déterminisme de l'effet terroir et gestion oenologique en Val de Loire. Application aux vins rouges de Cabernet franc et aux vins blancs moelleux de Chenin. Revue Francaise d'Oenologie, Paris 156, 14–20.

Balling, R.C. (1988) The climatic impact of a Sonoran vegetation discontinuity. Climatic Change 13, 99–109.

Balling, R.C. (1991) Impact of desertification on regional and global warming. Bulletin of the American Meteorological Society 72, 232–234.

Balling, R.C., Idso, S.B. and Hughes, W.S. (1992) Long-term and recent anomalous temperature changes in Australia. Geophysical Research Letters 19, 2317–2320.

Balling, R.C., Klopatec, J.M., Hildebrandt, M.L., Moritz, C.K. and Watts, C.J. (1998) Impacts of land degradation on historical temperature records from the Sonoran desert. Climatic Change 40, 669–681.

Ban, T., Ishimaru, M., Kobayashi, S., Shiozaki, S., Goto-Yamamoto, N. and Horiuchi, S. (2003) Abscisic acid and 2,4-dichlorophenoxyacetic acid affect the expression of anthocyanin biosynthetic pathway genes in 'Kyoho' grape berries. Journal of Horticultural Science & Biotechnology 78, 586–589.

Barbeau, G., Asselin, C. and Morlat, R. (1998) Estimation du potential viticole des terroirs en Val de Loire selon un indice de précocité du cycle de la vigne. Bulletin de l'O.I.V. 71, 247–262.

Barber, S.A. (1968) Mechanism of potassium absorption by plants. In 'The Role of Potassium in Agriculture', ed. V.J. Kilmer, S.E. Younts and N.C. Brady, 309–324. American Society of Agronomy et al., Madison.

Barton, H. (1991) The wines and vines of England. Grower [England], 16 May 1991, 11–16.

Bates, T.R., Dunst, R.M. and Joy, P. (2002) Seasonal dry matter, starch, and nutrient distribution in 'Concord' grapevine roots. HortScience 37, 313–316.

Bavaresco, L. (1989) Nutrizione minerale e resistenza alle malattie (dovute a fattori biotici) della vite. Vignevini 9, 25–35.

Becker, N.J. (1977) Selection of vineyard sites in cool climates. Proceedings, Third Australian Wine Industry Technical Conference, Albury 1977, 25–30.

Becker, N.J. (1988) Wine in the history of civilization and its position in modern society. Proceedings, Second International Symposium for Cool Climate Viticulture and Oenology, Auckland 1988, 289–293.

Beling, E., Sparks, A. and Tunstall, B. (2001) Prediction of vine development across Australia. Australian Grapegrower and Winemaker No. 450, 18–24.

Bell, S.-J. and Henschke, P.A. (2005) Implications of nitrogen nutrition for grapes, fermentation and wine. Australian Journal of Grape and Wine Research 11, 242–295.

Bellincontro, A., Fardelli, A., de Santis, D., Botondi, R. and Mencarelli, F. (2006) Postharvest ethylene and 1-MCP treatments both affect phenols, anthocyanins, and aromatic quality of Aleatico grapes and wine. Australian Journal of Grape and Wine Research 12, 141–149.

Bergqvist, J., Dokoozlian, N. and Ebisuda, N. (2001) Sunlight exposure and temperature effects on berry growth and composition of Cabernet Sauvignon and Grenache in the Central San Joaquin Valley of California. American Journal of Enology and Viticulture 52, 1–7.

Bernard, M., Horne, P.A. and 10 others (2007) Guidelines for environmentally sustainable winegrape production in Australia: IPM adoption self-assessment guide for growers. Australian and New Zealand Grapegrower and Winemaker No. 518, 24–35.

Bisson, L.F. (1991) Influence of nitrogen on yeast and fermentation of grapes. Proceedings, International Symposium on Nitrogen in Grapes and Wine, Seattle 1991, 78–89.

Böhm, R. (1998) Urban bias in temperature time series – a case study for the city of Vienna, Austria. Climatic Change 38, 113–128.

BOMA (Bureau of Meteorology, Australia) (2000) Climate data summaries. Private communication 2000.

Bonan, G.B. (1997) Effects of land use on the climate of the United States. Climatic Change 37, 449–486.

Bonan, G.B. (2001) Observational evidence for reduction of daily maximum temperature by croplands in the midwest United States. Journal of Climate 14, 2430–2442.

Bond, G., Kromer, B. and 8 others (2001) Persistent solar influence on North Atlantic climate during the Holocene. Science 294, 2130–2136.

Bondada, B.R., Matthews, M.A. and Shackel, K.A. (2005) Functional xylem in the post-veraison grape berry. Journal of Experimental Botany 56, 2949–2957.

Bootsma, A. (1976) Estimating minimum temperature and climatological freeze risk in hilly terrain. Agricultural Meteorology 16, 425–443.

Boulton, R. (1980) The general relationship between potassium, sodium and pH in grape juice and wine. American Journal of Enology and Viticulture 31, 182–186.

Boulton, R. (2001) The copigmentation of anthocyanins and its role in the colour of red wine: a critical review. American Journal of Enology and Viticulture 52, 67–87.

Bouma, T.J., de Visser, R., van Leeuwen, P.H., de Kock M.J. and Lambers, H. (1995) The respiratory energy requirements involved in nocturnal carbohydrate export from starch-storing mature source leaves and their contribution to leaf dark respiration. Journal of Experimental Botany 46, 1185–1194.

Boursiquot, J.M., Dessup, M. and Rennes, C. (1995) Distribution des principaux caractères phénologiques, agronomiques et technologiques chez Vitis vinifera L. Vitis 34(1), 31–35.

Bradley, R.S. (1991) Pre-instrumental climate: how has climate varied during the past 500 years? In 'Greenhouse-Gas-Induced Climatic Change: A Critical Appraisal of Simulations and Observations', ed. M.E. Schlesinger, 391–410. Elsevier, Amsterdam.

Bradley, R.S. (2000) Past global changes and their significance for the future. Quaternary Science Reviews 19, 391–402.

Branas, J. (1946) In Branas, J., Bernon, G and Levadoux, L.: Eléments de Viticulture Générale. Montpellier.

Branas, J. (1974) Viticulture. Déhan, Montpellier.

Briffa, K.R., Bartholin,T.S., Eckstein,D., Jones, P.D., Karlén, W., Schweingruber, F.H. and Zetterberg, P. (1990) A 1,400-year tree-ring record of summer temperatures in Fennoscandia. Nature 346, 434–439.

Briffa, K.R., Jones, P.D., Schweingruber, F.H. and Osborn, T.J. (1998a) Influence of volcanic eruptions on Northern Hemisphere summer temperature over the last 600 years. Nature 393, 450–454.

Briffa, K.R., Schweingruber, F.H., Jones, P.D., Osborn, T.J., Harris, I.C., Shiyatov, S.G., Vaganov, E.A. and Grudd, H. (1998b) Trees tell of past climates: but are they speaking less clearly today? Philosophical Transactions of the Royal Society of London B, 353, 65–73.

Briffa, K.R., Schweingruber, F.H., Jones, P.D., Osborn, T.J., Shiyatov, S.G. and Vaganov, E.A. (1998c) Reduced sensitivity of recent tree-growth to temperature at high northern latitudes. Nature 391, 678–682.

Briffa, K.R., Osborn, T.J., Schweingruber, F.H., Harris, I.C., Jones, P.D., Shiyatov, S.G. and Vaganov, E.A. (2001) Low-frequency temperature variations from a northern tree ring density network. Journal of Geophysical Research 106 (D3), 2929–2941.

Brohan, P., Kennedy, J.J., Harris, I., Tett, S.F.B. and Jones, P.D. (2006) Uncertainty estimates in regional and global observed temperature changes: a new data set from 1850. Journal of Geophysical Research 111, D12106, doi:10.1029/2005JD006548.

Brossaud, F., Cheynier, V., Asselin, C. and Moutounet, M. (1999) Flavonoid compositional differences of grapes among site test plantings of Cabernet Franc. American Journal of Enology and Viticulture 50, 277–284.

Bruer, D. (2001) Pruning, training and trellising in an organic vineyard. Australian Grapegrower and Winemaker No. 448, 31–32.

Buckerfield, J. and Webster, K. (2000) Vineyard trials show value of mulches –organic matter for management of young vines. Australian Grapegrower and Winemaker No. 441, 33–39.

Buckerfield, J. and Webster, K. (2001a) Managing earthworms in vineyards – improve incorporation of lime and gypsum. Australian Grapegrower and Winemaker No. 449a (Annual Technical Issue), 55–61.

Buckerfield, J. and Webster, K. (2001b) Responses to mulch continue – results from five years of field trials. Australian Grapegrower and Winemaker No. 453, 71–78.

Buckerfield, J. and Webster, K. (2002) Organic matter management in vineyards – mulches for soil maintenance. Australian and New Zealand Grapegrower and Winemaker No. 461, 26–33.

Busby, J. (1825) A Treatise on the Culture of the Vine and the Art of Making Wine. Government Printer, Sydney. (Facsimile reprint 1979, David Ell Press, Sydney.)

Busby, J. (1833) Journal of a Tour through some of the Vineyards of Spain and France. Stephens and Stokes, Sydney. (Facsimile reprint 1979, David Ell Press, Sydney.)

Buttrose, M.S. (1969) Vegetative growth of grapevine varieties under controlled temperature and light intensity. Vitis 8, 280–285.

Buttrose, M.S. and Hale, C.R. (1973) Effect of temperature on development of the grapevine inflorescence after bud burst. American Journal of Enology and Viticulture 24, 14–16.

Cacho, J., Castells, J.E., Esteban, A., Laguna, B. and Sagrista, N. (1995) Iron, copper and manganese influence on wine oxidation. American Journal of Enology and Viticulture 46, 380–384.

Campbell, C.D. and Sage, R.F. (2006) Interactions between the effects of atmospheric CO_2 content and P nutrition on photosynthesis in white lupin (*Lupinus albus* L.). Plant, Cell and Environment 29, 844–853.

Cane, M.A., Clement, A.C., Kaplan, A., Kushnir, Y., Pozdnyakov, D., Seager, R., Zebiak, S.E. and Murtugudde, R. (1997) Twentieth-century sea surface temperature trends. Science 275, 957–960.

Canny, M.J. (1973) Phloem Translocation. Cambridge University Press, London.

Cashman, G. (2000) Advection frost: a case study of the freeze of 27–28 October 1998 and its effects on the vineyards of the Canberra district. Part 1. Meteorological aspects. Australian Grapegrower and Winemaker No. 438, 74–84.

Cass, A. (2005) Effects of soil physical characteristics on mineral nutrient availability, movement and uptake. Proceedings, symposium 'Soil Environment and Mineral Nutrition', 3–11. American Society for Enology and Viticulture, Davis. (Cited by Cass and Roberts 2006.)

Cass, A. and Roberts, D. (2006) Poor soil physical properties can hinder nutrient uptake in vines. Australian and New Zealand Grapegrower and Winemaker No. 509a (Annual Technical Issue), 19–22.

Cass, A., Maschmedt, D. and Myburgh, P. (1998) Soil structure – are there best practices? Proceedings, seminar 'Viticultural Best Practice', Mildura 1997, 40–46. Australian Society of Viticulture and Oenology, Adelaide.

Cawthon, D.L. and Morris, J.R. (1982) Relationship of seed number and maturity to berry development, fruit maturation, hormonal changes and uneven ripening of 'Concord' (*Vitis labrusca* L.) grapes. Journal of the American Society for Horticultural Science 107, 1097–1104.

Cess, R.D., Potter, G.L. and 18 others (1989) Interpretation of cloud–climate feedback as produced by 14 atmospheric general circulation models. Science 245, 513–516.

Cess, R.D., Zhang, M.H. and 32 others (1996) Cloud feedback in atmospheric general circulation models: an update. Journal of Geophysical Research 101 (D8), 12791–12794.

Champagnol, F. (1984) Eléments de Physiologie de la Vigne et de Viticulture Générale. F. Champagnol, St-Gély-du-Fesc.

Chardonnet, C. and Donèche, B. (1995) Relation entre la teneur en calcium et la résistance à la digestion enzymatique du tissue pelliculaire au cours de la maturation du raisin. Vitis 34, 95–98.

Charlock, T.P. (1982) Cloud optical feedback and climate stability in a radiative-convective model. Tellus 34, 245–254.

Charlson, R.J., Langner, J., Rodhe, H., Leovy, C.B. and Warren, S.G. (1991) Perturbation of the Northern Hemisphere radiative balance by backscattering from anthropogenic sulphate aerosols. Tellus 43AB, 152–163.

Charlson, R.J., Schwartz, S.E., Hales, J.M., Cess, R.D., Coakley, J.A., Hansen, J.E. and Hofmann, D.J. (1992) Climate forcing by anthropogenic aerosols. Science 255, 423–430.

Cheng, L., Xia, G. and Bates, T. (2004) Growth and fruiting of young 'Concord' grapevines in relation to reserve nitrogen and carbohydrates. Journal of the American Society for Horticultural Science 129, 660–666.

Choné, X. (2003) Terroir et potential aromatique du Sauvignon Blanc: Acquisitions récentes. Progrès Agricole et Viticole 120, 63–68. (Cited by Bell and Henschke 2005.)

Choné, X., van Leeuwen, C., Chéry, P. and Ribéreau-Gayon, P. (2001) Terroir influence on water status and nitrogen status of non-irrigated Cabernet Sauvignon (*Vitis vinifera*). Vegetative development, must and wine composition (example of a Médoc top estate vineyard, Saint Julien area, Bordeaux 1997). South African Journal of Enology and Viticulture 22, 8–15.

Cocks, P.S. (1973) The influence of temperature and density on the growth of communities of subterranean clover (*Trifolium subterraneum* L. cv. Mount Barker). Australian Journal of Agricultural Research 24, 479–495.

Collander, R. (1959) Cell membranes: their resistance to penetration and their capacity for transport. In 'Plant Physiology: a Treatise', ed. F.C. Steward, Vol. II, 3–102. Academic Press, New York.

Collins, M.J., Barlow, E.W.R., Wood, R., Kelley, G. and Fuentes, S. (2005) Physiological, growth, yield and quality responses of 'Shiraz' berries manipulated using PRD and drip irrigation. Acta Horticulturae 689, 365–372.

Comrie, A.C. (2000) Mapping a wind-modified heat island in Tucson, Arizona (with comments on integrating research and undergraduate learning). Bulletin of the American Meteorological Society 81, 2417–2431.

Conradie, W.J. (1991) Translocation and storage of nitrogen by grapevines as affected by time of application. Proceedings, International Symposium on Nitrogen in Grapes and Wine, Seattle 1991, 32–42.

Conroy, J.P., Milham, P.J., Reed, M.L. and Barlow, E.W. (1990) Increases in phosphorus requirements for CO_2-enriched pine species. Plant Physiology 92, 977–982.

Cook, E.R., Palmer, J.G. and D'Arrigo, R.D. (2002) Evidence for a 'Mediaeval Warm Period' in a 1,100 year tree-ring reconstruction of past austral summer temperatures in New Zealand. Geophysical Research Letters 29 No. 14, 10.1029/2001GL014580.

Cook, E.R., Woodhouse, C.A., Eakin, C.M., Meko, D.M. and Stahle, D.W. (2004) Long-term aridity changes in the western United States. Science 306, 1015–1018.

Coombe, B.G. (1960) The relationship of growth and development to changes in sugars, auxins, and gibberellins in fruit of seeded and seedless varieties of Vitis vinifera. Plant Physiology 35, 241–250.

Coombe, B.G. (1973) The regulation of set and development of the grape berry. Acta Horticulturae 34, 261–273.

Coombe, B.G. (1976) The development of fleshy fruits. Annual Review of Plant Physiology 27, 507–528.

Coombe, B.G. (1988) Grapevine phenology. In 'Viticulture Volume 1 – Resources', first edition, ed. B.G. Coombe and P.R. Dry, 139–153. Australian Industrial Publishers, Adelaide.

Coombe, B.G. (2000) Hen and chicken vs sweet and sour. Australian and New Zealand Wine Industry Journal 15 (2), 13.

Coombe, B.G. and Hale, C.R. (1973) The hormone content of ripening grape berries and the effects of growth substance treatments. Plant Physiology 51, 629–634.

Coombe, B.G. and Iland, P.G. (2004) Winegrape quality. In 'Viticulture Volume 1 – Resources', second edition, ed. P.R. Dry and B.G. Coombe, 210–248. Winetitles, Adelaide.

Coombe, B.G. and McCarthy, M.G. (1997) Identification and naming of the inception of aroma development in ripening grape berries. Australian Journal of Grape and Wine Research 3, 18–20.

Coombe, B.G. and McCarthy, M.G. (2000) Dynamics of grape berry growth and physiology of ripening. Australian Journal of Grape and Wine Research 6, 131–135.

Cooper, M. (1993) The Wines and Vineyards of New Zealand, fourth edition, Hodder & Stoughton, Auckland.

Coquand, H. (1858) Description physique, géologique, paléontologique et minéralogique de la Charente. (Cited by Wilson 1998 and Galet 2000.)

Couzin, J. (1999) Landscape changes make regional climate run hot and cold. Science 283, 317–318.

Dai, A., Trenberth, K.E. and Karl, T.R. (1999) Effects of clouds, soil moisture, precipitation, and water vapour on diurnal temperature range. Journal of Climate 12, 2451–2473.

Daurel, J. (1892) Les Raisins de cuve de la Gironde et du Sud-ouest de la France. Bordeaux. (Cited by Robinson 2006.)

Debuigne, G. (1976) Larousse Dictionary of Wines of the World. Hamlyn, London.

Delas, J., Molot, C. and Soyer, J.P. (1991) Effects of nitrogen fertilization and grafting on the yield and quality of the crop of *Vitis vinifera* cv. Merlot. Proceedings, International Symposium on Nitrogen in Grapes and Wine, Seattle 1991, 242–248.

Deytieux, C., Gagné, S., L'Hyvernay, A., Donèche, B. and Geny, L. (2007) Possible roles of both abscisic acid and indol-acetic acid in controlling grape berry ripening process. Journal International des Sciences de la Vigne et du Vin 41, 141–148. (Cited by Antolin et al. 2008.)

Downey, M.O., Harvey, J.S. and Robinson, S.P. (2003a) Analysis of tannins in seeds and skins of Shiraz grapes throughout berry development. Australian Journal of Grape and Wine Research 9, 15–27.

Downey, M.O., Harvey, J.S. and Robinson, S.P. (2003b) Synthesis of flavonols and expression of flavonol synthase genes in the developing grape berries of Shiraz and Chardonnay (*Vitis vinifera* L.). Australian Journal of Grape and Wine Research 9, 110–121.

Downton, W.J.S. and Loveys, B.R. (1978) Compositional changes during grape berry development in relation to abscisic acid and salinity. Australian Journal of Plant Physiology 5, 415–423.

Downton, W.J.S., Björkman, O. and Pike, C.S. (1980) Consequences of increased atmospheric concentrations of carbon dioxide for growth and photosynthesis of higher plants. In 'Carbon Dioxide and Climate: Australian Research', ed. G.I. Pearman, 143–151. Australian Academy of Science, Canberra.

Dreier, L.P., Stoll, G.S. and Ruffner, H.P. (2000) Berry ripening and evapotranspiration of *Vitis vinifera* L. American Journal of Enology and Viticulture 51, 340–346.

Dry, P.R. and Botting, D.G. (1993) The effect of wind on the performance of Cabernet Franc grapevines. 1. Shoot growth and fruit yield components. Australian and New Zealand Wine Industry Journal 8, 347–352.

Dry, P.R. and Gregory, G.R. (1988) Grapevine varieties. In 'Viticulture Volume 1 – Resources', first edition, ed. B.G. Coombe and P.R. Dry, 119–138. Australian Industrial Publishers, Adelaide.

Dry, P.R. and Loveys, B.R. (1998) Factors influencing grapevine vigour and the potential for control with partial rootzone drying. Australian Journal of Grape and Wine Research 4, 140–148.

Dry, P.R., Loveys, B.R., Botting, D.G. and Düring, H. (1996) Effects of partial root-zone drying on grapevine vigour, yield, composition of fruit and use of water. Proceedings, Ninth Australian Wine Industry Technical Conference, Adelaide 1995, 128–131.

Dry, P.R., Loveys, B.R., Iland, P.G., Botting, D.G., McCarthy, M.G. and Stoll, M. (1999) Vine manipulation to meet fruit specifications. Proceedings, Tenth Australian Wine Industry Technical Conference, Sydney 1998, 208–214.

Dry, P.R., Loveys, B.R., Stoll, M., Steward, D. and McCarthy, M.G. (2000) Partial root-zone drying – an update. Australian Grapegrower and Winemaker No. 438a (Annual Technical Issue), 35–39.

Dry, P., Loveys, B., Stoll, M. and McCarthy, M. (2002) Commercial implementation of partial rootzone drying. Australian and New Zealand Wine Industry Journal 17(1), 46–48.

Dubourdieu, D. (1990) Complementarity of grape varieties and their sensory influence in the style of red and white great Bordeaux wines. Proceedings, Seventh Australian Wine Industry Technical Conference, Adelaide 1989, 5–7.

Due, G. (1994) Climatic effects are less important than site and year in modelling flowering and harvest dates. Australian and New Zealand Wine Industry Journal 9(1), 56–57.

Due, G. (1995) Continuing the climate debate. Australian and New Zealand Wine Industry Journal 10(1), 17–18.

Due, G., Morris, M., Pattison, S. and Coombe, B.G. (1993) Modelling grapevine phenology against weather considerations based on a large data set. Agricultural and Forest Meteorology 65, 91–106.

Dukes, B., Goldspink, B., Elliott, J. and Frayne, R. (1991) Time of nitrogenous fertilization can reduce fermentation time and improve wine quality. Proceedings, International Symposium on Nitrogen in Grapes and Wine, Seattle 1991, 249–254.

Easterling, D.R., Horton, B. and 9 others (1997) Maximum and minimum temperature trends for the globe. Science 277, 364–367.

Ebadi, A., May, P. and Coombe, B.G. (1996) Effect of short-term temperature and shading on fruit-set, seed and berry development in model vines of V. vinifera, cvs Chardonnay and Shiraz. Australian Journal of Grape and Wine Research 2, 2–9.

Eddy, J.A. (1976) The Maunder Minimum. Science 192, 1189–1202.

Eddy, J.A. (1977) Climate and the changing sun. Climatic Change 1, 173–190.

Edson, C.E., Howell, G.S. and Flore, J.A. (1995) Influence of crop load on photosynthesis and dry matter partitioning of Seyval grapevines. III. Seasonal changes in dry matter partitioning, vine morphology, yield, and fruit composition. American Journal of Enology and Viticulture 46, 478–482.

Engardt, M. and Rodhe, H. (1993) A comparison between patterns of temperature trends and sulfate aerosol pollution. Geophysical Research Letters 20, 117–120.

Esper, J., Cook, E.R. and Schweingruber, F.H. (2002) Low-frequency signals in long tree-ring chronologies for reconstructing past temperature variability. Science 295, 2250–2253.

Fairbridge, R.W. (1962) World sea-level and climatic changes. Quaternaria 6, 111–134.

Folland, C.K. and Parker, D.E. (1995) Correction of instrumental biases in historical sea surface temperature data. Quarterly Journal of the Royal Meteorological Society 121, 319–367.

Folland, C.K., Parker, D.E. and Kates, F.E. (1984) Worldwide marine temperature fluctuations 1856–1981. Nature 310, 670–673.

Fondy, B.R. and Geiger, D.R. (1982) Diurnal pattern of translocation and carbohydrate metabolism in source leaves of Beta vulgaris L. Plant Physiology 70, 671–676.

Fondy, B.R. and Geiger, D.R. (1985) Diurnal changes in allocation of newly fixed carbon in exporting sugar beet leaves. Plant Physiology 78, 753–757.

Foukal, P. and Lean, J. (1990) An empirical model of total solar irradiance variation between 1874 and 1988. Science 247, 556–558.

Frauenfeld, O.W., Davis, R.E. and Mann, M.E. (2005) A distinctly interdecadal signal of Pacific Ocean–Atmosphere interaction. Journal of Climate 18, 1709–1718.

Fregoni, M. (1977) Effects of the soil and water on the quality of the harvest. Proceedings, International Symposium on the Quality of the Vintage, Cape Town 1977, 151–168.

Friis-Christensen, E. and Lassen, K. (1991) Length of the solar cycle: an indicator of solar activity closely associated with climate. Science 254, 698–700.

Galet, P. (1958, 1962) Cépages et Vignobles de France, Tomes 2 and 3. Paul Déhan, Montpellier.

Galet, P. (1979) A Practical Ampelography. Trans. from the French by Lucie Morton. Cornell University Press, Ithaca and London.

Galet, P. (2000) General Viticulture. Trans. from the French by J. Towey. Oenoplurimédia, Chaintré.

Gallo, K.P., Adegoke, J.O., Owen, T.W. and Elvidge, C.D. (2002) Satellite-based detection of global urban heat-island temperature influence. Journal of Geophysical Research 107, D24, 4776, doi:10.1029/2002JD002588.

Garnier, M. (1967) Climatologie de la France: Selection de Données Statistiques. Mémorial de la Météorologie, Paris.

Gawel, R. (1998) Red wine astringency: a review. Australian Journal of Grape and Wine Research 4, 74–95.

Geiger, R. (1966) The Climate near the Ground. English translation by Scripta Technica Inc. of the 4th German edition. Harvard University Press, Cambridge, Mass.

Gershenzon, J. and Croteau, R.B. (1993) Terpenoid biosynthesis: the basic pathway and formation of monoterpenes, sesquiterpenes, and diterpenes. In 'Lipid Metabolism in Plants', ed. T.S. Moore, 339–388. CRC Press, Boca Raton.

Gladstones, J. (1992) Viticulture and Environment. Winetitles, Adelaide.

Gladstones, J. (1994) Climatology of cool viticultural environments: reply to criticism and a further examination of Tasmania and New Zealand. Australian and New Zealand Wine Industry Journal 9 (4), 349–361.

Gladstones, J. (1996) The Due model – the last word. Australian and New Zealand Wine Industry Journal 11(2) 134–139.

Gladstones, J. (2004) Climate and Australian viticulture. In 'Viticulture Volume 1 – Resources', second edition, ed. P.R. Dry and B.G. Coombe, 90–118. Winetitles, Adelaide.

Goldberg, D.M., Tsang, E., Karumanchiri, A. and Soleas, G.J. (1998) Quercetin and p-coumaric acid concentrations in commercial wines. American Journal of Enology and Viticulture 49, 142–151.

Goldspink, B. and Frayne, R. (1996) The effect of nutrients on vine performance, juice parameters and fermentation characteristics. Proceedings, seminar 'Quality Management in Viticulture', Mildura 1996, 17–21. Australian Society of Viticulture and Oenology, Adelaide.

Gornitz, V., Lebedeff, S. and Hansen, J. (1982) Global sea level trend in the past century. Science 215, 1611–1614.

Govindasamy, B., Duffy, P.B. and Caldeira, K. (2001) Land use changes and Northern Hemisphere cooling. Geophysical Research Letters 28, 291–294.

Gray, J.D. and Coombe, B.G. (2009) Variation in Shiraz berry size originates before fruitset but harvest is a point of resynchronisation for berry development after flowering. Australian Journal of Grape and Wine Research 15, 156–165.

Greenough, J.D., Longerich, H.P. and Jackson, S.E. (1997) Element fingerprinting of Okanagan Valley wines using ICP-MS: Relationships between wine composition, vineyard and wine colour. Australian Journal of Grape and Wine Research 3, 75–83.

Greer, D.H. and La Borde, D. (2006) Sunburn of grapes affects wine quality. Australian and New Zealand Grapegrower and Winemaker No. 506, 21–23.

Grifoni, D., Mancini, M., Maracci, G., Orlandini, S. and Zipoli, G. (2006) Analysis of Italian wine quality using freely available meteorological information. American Journal of Enology and Viticulture 57, 339–346.

Grimmer, C., Bachfischer, T. and Komor, E. (1999) Carbohydrate partitioning into starch in leaves of *Ricinus communis* L. grown under elevated CO_2 is controlled by sucrose. Plant, Cell and Environment 22, 1275–1280.

Grimstad, S.O. and Frimanslund, E. (1993) Effect of different day and night temperature regimes on greenhouse cucumber young plant production, flower bud formation and early yield. Scientia Horticulturae 53, 191–204.

Gu, S., Du, G., Zoldoske, D., Hakim, A., Cochran, R., Fugelsang, K. and Jorgensen, G. (2004) Effects of irrigation amount on water relations, vegetative growth, yield and fruit composition of Sauvignon blanc grapevines under partial rootzone drying and conventional irrigation in the San Joaquin Valley of California, USA. Journal of Horticultural Science and Biotechnology 79, 26–33.

Gu, S., Jacobs, S., Du, G., Guan, X. and Wample, R. (2008) Efficacy of ABA application to enhance fruit color of Cabernet Sauvignon grape in a warmer growing region. American Journal of Enology and Viticulture 59, 338A. (Abstract from ASEV 59th Annual Meeting 2008.)

Haigh, J.D. (1994) The role of stratospheric ozone in modulating the solar radiative forcing of climate. Nature 370, 544–546.

Haigh, J.D. (1996) The impact of solar variability on climate. Science 272, 981–984.

Haigh, J.D. (2003) The effects of solar variability on the Earth's climate. Philosophical Transactions of the Royal Society of London A, 361, 95–111.

Hale, C.R. (1968) Growth and senescence of the grape berry. Australian Journal of Agricultural Research 19, 939–945.

Halliday, J. (1991) Wine Atlas of Australia and New Zealand. Angus & Robertson, Sydney.

Halliday, J. (1993) Wine Atlas of California. Angus & Robertson, Sydney.

Halliday, J. (2006) Wine Atlas of Australia. Hardie Grant Books, Prahran, Victoria.

Hamman, R.A., Dami, I.-E., Walsh, T.M. and Stushnoff, C. (1996) Seasonal carbohydrate changes and cold hardiness of Chardonnay and Riesling grapevines. American Journal of Enology and Viticulture 47, 31–36.

Hancock, J.M. and Price, M. (1990) Real chalk balances the water supply. Journal of Wine Research 1(1), 45–60.

Hansen, J. and Lebedeff, S. (1987) Global trends of measured surface air temperature. Journal of Geophysical Research 92, 13345–13372.

Hansen, J., Johnson, D., Lacis, A., Lebedeff, S., Lee, P., Rind, D. and Russell, G. (1981) Climate impact of increasing atmospheric carbon dioxide. Science 213, 957–966.

Hansen, J., Sato, M., Lacis, A. and Ruedy, R. (1997) The missing climate forcing. Philosophical Transactions of the Royal Society of London B, 352, 231–240.

Hansen, J., Ruedy, R., Sato, M., Imhoff, M., Lawrence, W., Easterling, D., Peterson, T. and Karl, T. (2001) A closer look at United States and global surface temperature. Journal of Geophysical Research 106, D20, 23947–23963.

Hardie, W.J. (2000) Grapevine biology and adaptation to viticulture. Australian Journal of Grape and Wine Research 6, 74–81.

Harris, R.N. and Chapman, D.S. (2001) Mid-latitude (30°–60°N) climatic warming inferred by combining borehole temperatures with surface air temperatures. Geophysical Research Letters 28, 747–750.

Haselgrove, L., Botting, D., van Heeswijck, R., Høj, P.B., Dry P.R., Ford, C. and Iland, P.G. (2000) Canopy microclimate and berry composition: The effect of bunch exposure on the phenolic composition of Vitis vinifera L. cv. Shiraz grape berries. Australian Journal of Grape and Wine Research 6, 141–149.

Haubrick, L.L. and Assman, S.M. (2006) Brassinosteroids and plant function: some clues, more puzzles. Plant, Cell and Environment 29, 446–457.

Hegerl, G.C., Crowley, T.J., Allen, M., Hyde, W.T., Pollack, H.N., Smerdon J. and Zorita, E. (2007) Detection of human influence on a new, validated 1500-year temperature reconstruction. Journal of Climate 20, doi:10.1175/JCLI4011.1.

Henderson-Sellers, A. (1992) Continental cloudiness changes this century. Geological Journal 273, 255–262.

Henschke, P.A. and Jiranek, V. (1991) Hydrogen sulphide formation during fermentation: effect of nitrogen composition in model grape must. Proceedings, International Symposium on Nitrogen in Grapes and Wine, Seattle 1991, 172–183.

Hoffert, M.I. (1991) The effects of solar variability on climate. In 'Greenhouse-Gas-Induced Climatic Change: A Critical Appraisal of Simulations and Observations', ed. M.E. Schlesinger, 413–428. Elsevier, Amsterdam.

Hoitink, H.A.J., Wang, P. and Changa, C.M. (2002) Role of organic matter in plant health and soil quality. Proceedings, Eleventh Australian Wine Industry Technical Conference, Adelaide 2001, 57–60.

Howell, G.S. (2001) Grapevine cold hardiness: mechanisms of cold acclimation, mid-winter hardiness maintenance, and spring deacclimation. Proceedings, ASEV 50th Anniversary Annual Meeting, Seattle 2000, 35–48.

Hoyt, D.V. and Schatten, K.H. (1993) A discussion of plausible solar irradiance variations, 1700–1992. Journal of Geophysical Research 98(A ll), 18895–18906.

Hu, F.S., Kaufman, D. and 8 others (2003) Cyclic variation and solar forcing of holocene climate in the Alaskan subarctic. Science 301, 1890–1893.

Huang, S., Pollack, H.N. and Shen, P.-Y. (2000) Temperature trends over the past five centuries reconstructed from borehole temperatures. Nature 403, 756–758.

Huber, D.M. and Arny, D.C. (1985) Interactions of potassium with plant disease. In 'Potassium in Agriculture', ed. R.D. Munson, 467–488. American Society of Agronomy et al., Madison.

Huglin, P. (1978) Nouveau mode d'évaluation des possibilités héliothermiques d'un milieu viticole. C.R. Acad. Agr. France, 1117–1126. (Cited by Huglin 1983.)

Huglin, P. (1983) Possibilités d'appréciation objective du milieu viticole. Bulletin de l'O.I.V. 56, 823–833.

Huglin, P. (1986) Biologie et Ecologie de la Vigne. Payot Laussanne, Paris.

Hungate, B.A., Stiling, P.D., Dijkstra, P., Johnson, D.W., Ketterer, M.E., Hymus, G.J., Hinkle, C.R. and Drake, B.G. (2004) CO_2 elicits long-term decline in nitrogen fixation. Science 304, 1291.

Hyams, E. (1987) Dionysus: a Social History of the Wine Vine, second edition. Sidgwick & Jackson, London.

Iacono, F. and Sommer, K.J. (1996) Photoinhibition of photosynthesis and photorespiration in *Vitis vinifera* under field conditions: effects of light climate and leaf position. Australian Journal of Grape and Wine Research 2, 10–20.

IPCC (Intergovernmental Panel on Climate Change) (2007) Climate change 2007: the physical science basis. Summary for policymakers. IPCC, Geneva, February 2007, pp. 18.

Jackson, D.I. (1993a) Climate debate: Comments on *Viticulture and Environment* by J. Gladstones. Australian and New Zealand Wine Industry Journal 8(4), 1–2.

Jackson, D.I. (1993b) Degree days and the value of climatic indices. Australian and New Zealand Wine Industry Journal 8(4), 3.

Jackson, D.I. (1995a) Climate – a tale of two districts. Australian and New Zealand Wine Industry Journal 10(3), 226–229.

Jackson, D.I. (1995b) Cool-climate viticulture. Australian and New Zealand Wine Industry Journal 10(4), 362–365.

Jackson, D.I. (1996) Climate – LTI revisited. Australian and New Zealand Wine Industry Journal 11(1), 79–81.

Jackson, D.I. and Schuster, D. (2001) The Production of Grapes and Wine in Cool Climates, third edition. Brassell Assoc. and Gypsum Press, Wellington.

Jirikowic, J.L. and Damon, P.E. (1994) The mediaeval solar activity maximum. Climatic Change 26, 309–316.

Johnson, H. and Robinson, J. (2001) The World Atlas of Wine, fifth edition, Mitchell Beasley, London.

Joly, N. (1999) Wine from Sky to Earth: Growing and Appreciating Biodynamic Wine. Acres USA, Austin, Texas.

Jones, G.V. (2005) Climate change in the western United States grape growing regions. Acta Horticulturae 689, 41–59.

Jones, G.V. and Davis, R.E. (2000a) Using a synoptic climatological approach to understand climate–viticulture relationships. International Journal of Climatology 20, 813–837.

Jones, G.V. and Davis, R.E. (2000b) Climate influences on grapevine phenology, grape composition, and wine production and quality for Bordeaux, France. American Journal of Enology and Viticulture 51, 249–261.

Jones, G.V. and Goodrich, G.B. (2008) Influence of climate variability on wine regions in the western USA and on wine quality in the Napa Valley. Climate Research 35, 241–254.

Jones, G.V., White, M.A., Cooper, O.R. and Storchmann, K. (2005) Climate change and global wine quality. Climatic Change 73, 319–343.

Jones, P.A. and Henderson-Sellers, A. (1992) Historical records of cloudiness and sunshine in Australia. Journal of Climate 5, 260–267.

Jones, P.D. (1994) Hemispheric surface air temperature variations: a reanalysis and an update to 1993. Journal of Climate 7, 1794–1802.

Jones, P.D. and Mann, M.E. (2004) Climate over past millennia. Reviews of Geophysics 42, RG2002, 10.1029/2003RG000143.

Jones, P.D. and Moberg, A. (2003) Hemispheric and large-scale air temperature variations: an extensive revision and an update to 2001. Journal of Climate 16, 206–223.

Jones, P.D., Raper, S.C.B., Bradley, R.S., Diaz, H.F., Kelly, P.M and Wigley, T.M.L. (1986a) Northern Hemisphere surface air temperature variations: 1851–1984. Journal of Climate and Applied Meteorology 25, 161–179.

Jones, P.D., Raper, S.C.B. and Wigley, T.M.L. (1986b) Southern Hemisphere surface air temperature variations: 1851–1984. Journal of Climate and Applied Meteorology 25, 1213–1230.

Jones, P.D., Wigley, T.M.L. and Wright, P.B. (1986c) Global temperature variations between 1861 and 1984. Nature 322, 430–434.

Jones, P.D., Groisman, P.Y., Coughlan, M., Plummer, M., Wang, W.C. and Karl, T.R. (1990) Assessment of urbanization effects in the series of surface air temperature range over land. Nature 347, 169–172.

Jones, P.D., New, M., Parker, D.E., Martin, S. and Rigor, I.G. (1999) Surface air temperature and its changes over the past 150 years. Reviews of Geophysics 37, 173–199.

Jones, P.D., Briffa, K.R. and Osborn, T.J. (2003) Changes in the Northern Hemisphere annual cycle: implications for paleoclimatology? Journal of Geophysical Research 108, D18, 4588, doi:10.1029/2003JD003695.

Jordan, D. (1996) Magnesium for yield and quality. Australian Grapegrower and Winemaker No. 395, 55–56.

Jørgensen, T.S. and Hansen, A.W. (2000) Comments on 'Variation of cosmic ray flux and global cloud coverage – a missing link in solar–climate relationships' by Henrik Svensmark and Eigil Friis-Christensen. Journal of Atmospheric and Solar-Terrestrial Physics 62, 73–77.

Kaiser, D.P. (2000) Decreasing cloudiness over China: An updated analysis examining additional variables. Geophysical Research Letters 27, 2193–2196.

Kalnay, E. and Cai, M. (2003) Impact of urbanization and land-use change on climate. Nature 423, 528–531.

Kanellis, A.K. and Roubelakis-Angelakis, K.A. (1993) Grape. In 'Biochemistry of Fruit Ripening', ed. G. Seymour, J. Taylor and G. Tucker, 189–234. Chapman & Hall, London.

Kaplan, A., Cane, M.A., Kushnir, Y., Clement, A.C., Blumenthal, M.B. and Rajagopalan, B. (1998) Analyses of global sea surface temperature 1856–1991. Journal of Geophysical Research 103, C9, 18567–18589.

Karl, T.R. and Jones, P.D. (1989) Urban bias in area-averaged surface air temperature trends. Bulletin of the American Meteorological Society 70, 265–270.

Karl, T.R. and Jones, P.D. (1990) Reply [to G.M. Cohen]. Bulletin of the American Meteorological Society 71, 572–574.

Karl, T.R. and Steurer, P.M. (1990) Increased cloudiness in the United States during the first half of the twentieth century: fact of fiction? Geophysical Research Letters 17, 1925–1928.

Karl, T.R., Kukla, G. and Gavin, J. (1984) Decreasing diurnal temperature range in the United States and Canada from 1941 through 1980. Journal of Climate and Applied Meteorology 23, 1489–1504.

Karl, T.R., Kukla, G. and Gavin, J. (1986) Relationship between decreased temperature range and precipitation trends in the Unites States and Canada 1941–80. Journal of Climate and Applied Meteorology 25, 1878–1886.

Karl, T.R., Diaz, H.F. and Kukla, G. (1988) Urbanization: its detection and effect in the United States climate record. Journal of Climate 1, 1099–1123.

Karl, T.R., Kukla, G., Razuvayev, V.N., Changery, M.J., Quayle, R.G., Heim, R.R., Easterling, D.R. and Fu, C.B. (1991) Global warming: evidence for asymmetric diurnal temperature change. Geophysical Research Letters 18, 2253–2256.

Karl, T.R., Jones, P.D. and 8 others (1993) A new perspective on recent global warming: asymmetric trends of daily maximum and minimum temperature. Bulletin of the American Meteorological Society 74, 1007–1023.

Karlsson, M.G., Heins, R.D., Erwin, J.E., Berghage, R.D., Carlson, W.H. and Biernbaum, J.A. (1989) Temperature and photosynthetic photon flux influence chrysanthemum shoot development and flower initiation under short-day conditions. Journal of the American Society for Horticultural Science 114, 158–163.

Kaufman, Y.J. and Koren, I. (2006) Smoke and pollution aerosol effect on cloud cover. Science 313, 655–658.

Kelly, P.M. and Wigley, T.M.L. (1990) The influence of solar forcing trends on global mean temperature since 1861. Nature 347, 460–462.

Kelly, P.M. and Wigley, T.M.L. (1992) Solar cycle length, greenhouse forcing and global climate. Nature 360, 328–330.

Kennedy, J.A., Troup, G.J. and 7 others (2000) Development of polyphenols in berries from *Vitis vinifera* L. cv. Shiraz. Australian Journal of Grape and Wine Research 6, 244–254.

Kennedy, J.A., Matthews, M.A. and Waterhouse, A.L. (2002) Effect of maturity and vine water status on grape skin and wine flavonoids. American Journal of Enology and Viticulture 53, 268–274.

Kennedy, J.A., Ferrier, J., Harbertson, J.F. and Peyrot des Gachons, C. (2006) Analysis of tannins in red wine using multiple methods: correlation with perceived astringency. American Journal of Enology and Viticultre 57, 481–485.

Kernthaler, S.C., Toumi, R. and Haigh, J.D. (1999) Some doubts concerning a link between cosmic ray fluxes and global cloudiness. Geophysical Research Letters 26, 863–865.

Kerridge, G.H., Clingeleffer, P.R. and Possingham, J.V. (1987–1988) Varieties and varietal wines from the Merbein grape germplasm collection : 1. Varieties producing full bodied red wines. Australian Grapegrower and Winemaker No. 277, 14–18. 2. Varieties producing light bodied red wines. *Ibid* 279, 14–19. 3. Varieties producing aromatic white wines. *Ibid* 280, 29–34. 4. Varieties producing delicate white wines. *Ibid* 283, 17–23. 5. Varieties producing full bodied white wines. *Ibid* 292, 31–35.

Kiehl, J.T. and Briegleb, B.P. (1993) The relative roles of sulfate aerosols and greenhouse gases in climate forcing. Science 260, 311–314.

Kiehl, J.T. and Ramanathan, V. (1983) CO_2 radiative parameterization used in climate models: comparison with narrow band models and with laboratory data. Journal of Geophysical Research 88 (C9), 5191–5202.

Kiehl, J.T., Schneider, T.L., Portmann, R.W. and Solomon, S. (1999) Climate forcing due to tropospheric and stratospheric ozone. Journal of Geophysical Research 104 (D24), 31239–31254.

Kirk, J.T.O. and Hutchinson, M.F. (1994) Mapping of grapevine growing season temperature summation. Australian and New Zealand Wine Industry Journal 9, 247–251.

Koepf, H.H., Pettersson, B.D. and Schaumann, W. (1976) Bio-dynamic Agriculture.
Anthroposophic Press, Spring Valley, N.Y.

Kondo, S. and Tomiyama, A. (1998) Changes of free and conjugated ABA in the fruit of
'Satohnishiki' sweet cherry and the ABA metabolism after application of (s)-(+)-ABA.
Journal of Horticultural Science & Biotechnology 73, 467–472.

Kondo, S., Posuya, P., Kanlayanarat, S. and Hirai, N. (2001). Abscisic acid metabolism
during development and maturation of rambutan fruit. Journal of Horticultural
Science & Biotechnology 76, 235–241.

Kondo, S., Ponrod, W., Kanlayanarat, S. and Hirai, N. (2002) Abscisic metabolism during
fruit development and maturation of mangosteens. Journal of the American Society for
Horticultural Science 127, 737–741.

Kopec, R.J. (1967) Effects of the Great Lakes' thermal influence on freeze-free dates in
spring and fall as determined by Hopkins' bioclimatic law. Agricultural Meteorology
4, 241–253.

Körner, C. (2006) Plant CO_2 responses: an issue of definition, time and resource supply. New
Phytologist 172, 393–411.

Koussa, T., Broquedis, M. and Bouard, J. (1993) Mise en évidence d'une relation entre les
teneurs en acide abscissique des feuilles de vigne et des baies de raisin à l'époque de la
véraison. Journal International des Sciences de la Vigne et du Vin 27, 263–276.

Kriedemann, P.E., Törökfalvy, E. and Smart, R.E. (1973) Natural occurrence and photosynthetic utilisation of sunflecks by grapevine leaves. Photosynthetica 7, 18–27.

Kriedemann, P.E., Sward, R.J. and Downton, W.J.S. (1976) Vine response to carbon dioxide
enrichment during heat therapy. Australian Journal of Plant Physiology 3, 605–618.

Kristjánsson, J.E. and Kristiansen, J. (2000) Is there a cosmic ray signal in recent variations
in global cloudiness and cloud radiative forcing? Journal of Geophysical Research 105,
D9, 11851–11863.

Kristjánsson, J.E., Staple, A. and Kristiansen, J. (2002) A new look at possible connections
between solar activity, clouds and climate. Geophysical Research Letters 29, 23, 2107,
doi:10.1029/2002GL015646.

Kukla, G., Gavin, J. and Karl, T.R. (1986) Urban warming. Journal of Climate and Applied
Meteorology 25, 1265–1270.

Kunkee, R.E. (1991) Relationship between nitrogen content of must and sluggish fermentation. Proceedings, International Symposium on Nitrogen in Grapes and Wine, Seattle
1991, 148–155.

Laird, K.R., Fritz, S.C., Maasch, K.A. and Cumming, B.F. (1996) Greater drought intensity and frequency before AD 1200 in the northern Great Plains, USA. Nature 384,
552–554.

Lake, M. (1964) Hunter Wine. Jacaranda Press, Brisbane.

Lal, M. and Ramanathan, V. (1984) Effects of moist convection and water vapour radiative
processes on climate sensitivity. Journal of Atmospheric Sciences 41, 2238–2249.

La Marche, V.C., Graybill, D.A., Fritts, H.C. and Rose, M.R. (1984) Increasing atmospheric
carbon dioxide: tree ring evidence for growth enhancement in natural vegetation.
Science 225, 1019–1021.

Lamb, H.H. (1977) Climate: Present Past and Future. Volume 2: Climatic History and the Future. Methuen, London.

Lamb, H.H. (1982) Climate, History and the Modern World. Methuen, London.

Lamb, H.H. (1984) Climate of the last thousand years: natural climatic fluctuations and change. In 'The Climate of Europe: Past, Present and Future', ed. H. Flohn and R. Fantechi, 25–64. Reidel, Dordrecht.

Langridge, J. and McWilliam, J.R. (1967) Heat responses in higher plants. In 'Thermobiology', ed. A.H. Rose, 231–292. Academic Press, London.

Lassen, K. and Friis-Christensen, E. (1995) Variability of the solar cycle length during the past five centuries and the apparent association with terrestrial climate. Journal of Atmospheric and Terrestrial Physics 57, 835–845.

Laville, P. (1990) Le terroir, un concept indispensable à l'élaboration et à la protection des appellations d'origine comme à la gestation des vignobles: le cas de la France. Bulletin de l'O.I.V. 63, 217–241.

Lean, J. (2000) Evolution of the sun's spectral irradiance since the Maunder Minimum. Geophysical Research Letters 27, 2425–2428.

Lean, J. and Rind, D. (1998) Climate forcing by changing solar radiation. Journal of Climate 11, 3069–3094.

Lean, J., Beer, J. and Bradley, R. (1995) Reconstruction of solar irradiance since 1610: implications for climate change. Geophysical Research Letters 22, 3195–3198.

Lee, W.-H., Iacobellis, S.F. and Somerville, R.C.J. (1997) Cloud radiation forcings and feedbacks: general circulation model tests and observational validation. Journal of Climate 10, 2479–2496.

Lerdau, M., Litvak, M., Palmer, P. and Monson, R. (1997) Controls over monoterpene emissions from boreal forest conifers. Tree Physiology 17, 563–569.

Le Roy Ladurie, E. (1971) Times of Feast, Times of Famine. A History of Climate since the Year 1000. Allen and Unwin, London. (Trans. by Barbara Bray from the 1967 French edition.)

Le Roy Ladurie, E. and Baulant, M. (1980) Grape harvests from the fifteenth through the nineteenth centuries. Journal of Interdisciplinary History 10, 839–849. (Cited by Pfister 1988.)

Lockwood, M. (2001) Long-term variations in the magnetic fields of the Sun and the heliosphere: their origin, effects, and implications. Journal of Geophysical Research 106 (A8), 16021–16038.

Lockwood, M. and Stamper, R. (1999) Long-term drift of the coronal source magnetic flux and the total solar irradiance. Geophysical Research Letters 26, 2461–2464.

Lockwood, M., Stamper, R. and Wild, M.N. (1999) A doubling of the Sun's coronal magnetic field during the last 100 years. Nature 399, 437–439.

Loveys, B.R. (1984) Abscisic acid transport and metabolism in grapevine (*Vitis vinifera* L.) New Phytologist 98, 575–582.

Loveys, B.R. and Düring, H. (1984) Diurnal changes in the water relations and abscisic acid in field-grown *Vitis vinifera* cultivars. II. Abscisic acid changes under semi-arid conditions. New Phytologist 97, 37–47.

Loveys, B.R., Stoll, M., Dry, P.R. and McCarthy, M.G. (1998) Partial rootzone drying stimulates stress responses in grapevine to improve water use efficiency while maintaining crop yield and quality. Australian Grapegrower and Winemaker No. 414a (Annual Technical Issue), 108–113.

Loveys, B.R., Stoll, M. and Dry, P.R. (2001) Partial rootzone drying – how does it work? Australian Grapegrower and Winemaker No. 449a (Annual Technical Issue), 25–28.

Ludvigsen, K. (1995) Temperature of soil at three depths with straw mulch and bare soil treatments. Australian Grapegrower and Winemaker, Annual Technical Issue 1995, 103–109.

Luterbacher, J., Dietrich, D., Xoplaki, E., Grosjean, M. and Wanner, H. (2004) European seasonal and annual temperature variability, trends and extremes since 1500. Science 303, 1499–1503.

McCarthy, M.G. and Coombe, B.G. (1999) Is weight loss in ripening grape berries cv. Shiraz caused by impeded phloem transport? Australian Journal of Grape and Wine Research 5, 17–21.

McIntyre, G.N., Kliewer, W.M. and Lider, L.A. (1987) Some limitations of the degree day system as used in viticulture in California. American Journal of Enology and Viticulture 38, 128–132.

Mahrt, L. and Heald, R.C. (1993) Nocturnal surface temperature distribution as remotely sensed from low-flying aircraft. Agricultural Meteorology 28, 99–107.

Manabe, S. and Wetherald, R.T. (1967) Thermal equilibrium of the atmosphere with a given distribution of relative humidity. Journal of Atmospheric Sciences 24, 241–259.

Manabe, S. and Wetherald, R.T. (1975) The effects of doubling the CO_2 concentration on the climate of a general circulation model. Journal of Atmospheric Sciences 32, 3–15.

Mann, M.E., Bradley, R.S. and Hughes, M.K. (1998) Global-scale temperature patterns and climate forcing over the past six centuries. Nature 392, 779–787.

Mann, M.E., Bradley, R.S. and Hughes, M.K. (1999) Northern Hemisphere temperatures during the past millennium: Inferences, uncertainties, and limitations. Geophysical Research Letters 26, 759–762.

Mann, M.E., Rutherford, S., Bradley, R.S., Hughes, M.K. and Keimig, F.T. (2003) Optimal surface temperature reconstructions using terrestrial borehole data. Journal of Geophysical Research 108 (D7), 4203, doi:10.1029/2002JD002532.

Mannini, F., Calo, A. and Intrieri, C. (1997) Italian viticulture: focus on high quality native wine cultivars and their growing areas. Australian and New Zealand Wine Industry Journal 12 (4), 408–421.

Mantua, N.J. and Hare, S.R. (2002) The Pacific Decadal Oscillation. Journal of Oceanography 58, 35–44.

Marais, J., Calitz, F. and Haasbroek, P.D. (2001) Relationship between microclimatic data, aroma component concentrations and wine quality parameters in the prediction of Sauvignon Blanc wine quality. South African Journal for Enology and Viticulture 22, 22–26.

Marès, H. (1890) Description des Cépages Principaux de la Région Mediterranéenne de la France. Camille Coulet, Montpellier.

Marschner, H. (1995) Mineral Nutrition of Higher Plants, second edition, Academic Press, London.

Marsh, N.D. and Svensmark, H. (2000) Low cloud properties influenced by cosmic rays. Physical Review Letters 85, 5004–5007.

Maschmedt, D.J. (2004) Soils and Australian viticulture. In 'Viticulture Volume 1 – Resources', second edition, ed. P.R. Dry and B.G. Coombe, 56–89. Winetitles, Adelaide.

Maugeri, M., Bagnati, Z. and Brunetti, M. (2001) Trends in Italian total cloud amount, 1951–1996. Geophysical Research Letters 28, 4551–4554.

May, P. (2004) Flowering and Fruitset in Grapevines. Lythrum Press, Adelaide.

Mazliak, P. (1970) Lipids. In 'The Biochemistry of Fruits and their Products', ed. A.C. Hulme, Vol. 1, 209–238. Academic Press, London.

Meehl, G.A., Washington, W.M., Amman, C.M., Arblaster, J.M., Wigley, T.M.L. and Tebaldi, C. (2004) Combinations of natural and anthropogenic forcings in the twentieth-century climate. Journal of Climate 17, 3721–3727.

Meinert, L.D. and Busacca, A.J. (2000) Geology and wine. 3. Terroirs of the Walla Walla Valley appellation, southeastern Washington State, USA. Geoscience Canada 27 (4), 149–171.

Meinert, L.D. and Busacca, A.J. (2002) Geology and wine. 6. Terroir of the Red Mountain appellation, central Washington State, USA. Geoscience Canada 29(4), 149–168.

Millan, C., Vargas, A., Rubio, A., Moreno, J. and Ortega, J.M. (1992) Fatty acid content of unripe and ripe 'Pedro Ximenez' Vitis vinifera grapes. Journal of Wine Research 3, 235–240.

Miller, D.P., Howell, G.S. and Flore, J.A. (1997) Influence of shoot number and crop load on potted Chambourcin grapevines. II: Whole-vine vs. single-leaf photosynthesis. Vitis 36, 109–114.

Ministry of Agriculture, Fisheries and Food, France (1997) Catalogue of selected wine grape varieties and clones cultivated in France. Maraval, St Pons de Thomières.

Mitchell, J.F.B. and Johns, T.C. (1997) On modification of global warming by sulfate aerosols. Journal of Climate 10, 245–267.

Mitchell, J.F.B., Senior, C.A. and Ingram, W.J. (1989) CO_2 and climate: a missing feedback? Nature 341, 132–134.

Mitchell, J.F.B., Davis, R.A., Ingram, W.J. and Senior, C.A. (1995) On surface temperature, greenhouse gases, and aerosols: models and observations. Journal of Climate 8, 2364–2386.

Moberg, A., Sonechkin, D.M., Holmgren, K., Datsenko, N.M., Karlén, W. and Lauritzen, S.-E. (2005) Highly variable Northern Hemisphere temperatures reconstructed from low- and high-resolution proxy data. Nature 433, 613–617 and 439, 1014.

Morison, J.I.L. and Gifford, R.M. (1984) Plant growth and water use with limited water supply and high CO_2 concentrations. II. Plant dry weight, partitioning and water use efficiency. Australian Journal of Plant Physiology 11, 375–384.

Morlat, R. (2008) Long-term additions of organic amendments in a Loire Valley vineyard on a calcareous sandy soil. II. Effects on root system, growth, grape yield, and foliar nutrient status of a Cabernet franc vine. American Journal of Enology and Viticulture 59, 364–374.

Morlat, R. and Asselin, C. (1992) Un terroir de référence pour la qualité et la typicité des vins rouges du Val-de-Loire: la craie tuffeau. Bulletin de l'O.I.V. 65, 329–343.

Morlat, R. and Chaussod, R. (2008) Long-term additions of organic amendments in a Loire Valley vineyard. I. Effects on properties of a calcareous sandy soil. American Journal of Enology and Viticulture 59, 353–363.

Morlat, R. and Symoneaux, R. (2008) Long-term additions of organic amendments in a Loire Valley vineyard on a calcareous sandy soil. III. Effects on fruit composition and chemical and sensory characteristics of Cabernet franc wine. American Journal of Enology and Viticulture 59, 375–386.

Morlat, R., Jacquet, A. and Asselin, C. (1997) Variabilité de la précocité de la vigne en Val de Loire: rôle du terroir et du millésime, conséquences sur la composition de la baie. Revue Francaise d'Oenologie, Paris, 165, 11–22.

Morris, C.J.G., Simmonds, I. and Plummer, N. (2001) Quantification of the influences of wind and cloud on the nocturnal urban heat island of a large city. Journal of Applied Meteorology, 40, 169–182.

Mortensen, L.M. (1987) Review: CO_2 enrichment in greenhouses. Crop responses. Scientia Horticulturae 33, 1–25.

MOUK (Meteorological Office, United Kingdom) (1967) Tables of Temperature Relative Humidity and Precipitation for the World. Part III: Europe and the Atlantic Ocean north of 35°N, second edition. HMSO, London.

Mpelasoka, B.S., Schachtman, D.P., Treeby, M.T. and Thomas, M.R. (2003) A review of potassium nutrition in grapevines with special emphasis on berry accumulation. Australian Journal of Grape and Wine Research 9, 154–168.

Müller, M.J. (1982) Selected Climate Data for a Global Set of Standard Stations for Vegetation Science. Dr W. Junk, The Hague.

Mullins, M.G., Bouquet, A. and Williams, L.E. (1992) Biology of the Grapevine. Cambridge University Press, Cambridge.

Mundy, D.C. and Agnew, R.H. (2002) Do mulched vines produce better wine? Proceedings, Eleventh Australian Wine Industry Technical Conference, Adelaide 2001, 80–83.

Murat, M.-L. (2005) Recent findings on rosé wine aromas. Part 1: identifying aromas studying the aromatic potential of grapes and juice. Australian and New Zealand Grapegrower and Winemaker No. 497a (Annual Technical Issue), 64–76.

Murata, N. and Los, D.A. (1997) Membrane fluidity and temperature perception. Plant Physiology 115, 875–879.

Myster, J. and Moe, R. (1995) Effect of diurnal temperature alternations on plant morphology in some greenhouse crops – a mini review. Scientia Horticulturae 62, 205–215.

NCDC (National Climatic Data Center, USA) (1996) International Station Meteorological Climate Summary, Version 4.0, CD-ROM. NCDC, Asheville, NC.

NCDC (National Climatic Data Center, USA) (2002) The Climate Atlas of the USA, Version 2.0, CD-ROM. NCDC, Asheville, NC.

Nemani, R.R., White, M.A., Cayan, D.R., Jones, G.V., Running, S.W., Coughlan, J.C. and Peterson, D.L. (2001) Asymmetric warming over coastal California and its impact on the premium wine industry. Climate Research 19, 25–34.

NIWA (National Institute of Water and Atmospheric Research, New Zealand) (2003) Climate data summaries. Private communication, 10 April 2003.

Noble, A.C. (1979) Evaluation of Chardonnay wines obtained from sites with different soil compositions. American Journal of Enology and Viticulture 30 214–217.

Norris, J.R. (1999) On trends and possible artifacts in global ocean cloud cover between 1952 and 1995. Journal of Climate 12, 1864–1870.

Norton, R.A. (1979) Growing grapes for wine and table in the Puget Sound region. Extension Bulletin 0775, Washington State University, College of Agriculture, Pullman.

Oerlemans, J. (2005) Extracting a climate signal from 169 glacier records. Science 308, 675–677.

Osborn, T.J. and Briffa, K.R. (2004) The real colour of climate change? Science 306, 621–622.

Palejwala, V.A., Parikh, H.R. and Modi, V.V. (1985) The role of abscisic acid in the ripening of grapes. Physiologia Plantarum 65, 498–502.

Pan, Q.-H., Li, M.-J. and 8 others (2005) Abscisic acid activates acid invertases in developing grape berry. Physiologia Plantarum 125, 157–170.

Parker, D.E. (2004) Large-scale warming is not urban. Nature 432, 290.

Parungo, F., Boatman, J.F., Sievering, H., Wilkison, S.W. and Hicks, B.B. (1994) Trends in global marine cloudiness and anthropogenic sulfur. Journal of Climate 7, 434–440.

Pate, J.S., Emery, R.J.N. and Atkins, C.A. (1998) Transport physiology and partitioning. In 'Lupins as Crop Plants: Biology, Production and Utilization', ed. J.S. Gladstones, C.A. Atkins and J. Hamblin. CAB International, Wallingford.

Peterson, T.C. (2003) Assessment of urban versus rural in situ surface temperatures in the contiguous United States: no difference found. Journal of Climate 16, 2941–2959.

Peterson, T.C. and Owen, T.W. (2005) Urban heat island assessment: metadata are important. Journal of Climate 18, 2637–2646.

Peterson, T.C., Golubev, V.S. and Groisman, P.Ya. (1995) Evaporation losing its strength. Nature 377, 687–688.

Peterson, T.C., Gallo, K.P., Lawrimore, J., Owen, T.W., Huang, A. and McKittrick, D.A. (1999) Global rural temperature trends. Geophysical Research Letters 26, 329–332.

Petit-Lafitte, A. (1868) La Vigne dans le Bordelais. J. Rothschild, Paris.

Petrie, P.R. and Sadras, V.O. (2008) Advancement of grapevine maturity in Australia between 1993 and 2006: putative causes, magnitude of trends and viticultural consequences. Australian Journal of Grape and Wine Research 14, 33–45.

Petrie, P.R., Trought, M.C.T. and Howell, G.S. (2000a) Fruit composition and ripening of Pinot Noir (Vitis vinifera L.) in relation to leaf area. Australian Journal of Grape and Wine Research 6, 46–51.

Petrie, P.R., Trought, M.C.T. and Howell, G.S. (2000b) Influence of leaf ageing, leaf area and crop load on photosynthesis, stomatal conductance and senescence of grapevine (Vitis vinifera L. cv. Pinot noir) leaves. Vitis 39, 31–36.

Peynaud, E. and Ribéreau-Gayon, P. (1971) The grape. In 'The Biochemistry of Fruits and their Products', ed. A.C. Hulme, Vol. 2, 171–205. Academic Press, London.

Peyrot des Gachons, C., van Leeuwen, C., Tominaga, T., Soyer, J.-P., Gaudillère, J.-P. and Dubourdieu, D. (2005) Influence of water and nitrogen deficit on fruit ripening and aroma potential of Vitis vinifera L. cv. Sauvignon Blanc in field conditions. Journal of the Science of Food and Agriculture 85, 73–85.

Pfister, C. (1984) Klimageschichte der Schweiz 1525–1860. Das Klima der Schweiz von 1525–1860 und seine Bedeutung in der Geschichte von Bevölkerung und Landwirtschaft, Volume 1. Haupt, Bern. (Cited by Pfister 1988.)

Pfister, C. (1988) Variations in the spring–summer climate of central Europe from the High Middle Ages to 1850. In 'Long and Short Term Variability of Climate', ed. H. Wanner and U. Siegenthaler, 57–82. Springer-Verlag, Berlin.

Pierce, I. and Coombe, B.G. (2004) Grapevine phenology. In 'Viticulture Volume 1 – Resources', second edition, ed. P.R. Dry and B.G. Coombe, 150–166. Winetitles, Adelaide.

Pirie, A.J.G. and Mullins, M.G. (1976) Changes in anthocyanin and phenolics content of grapevine leaf and fruit tissues treated with sucrose, nitrate, and abscisic acid. Plant Physiology 58, 468–472.

Pitman, A.J., Narisma, G.T., Pielke, R.A. and Holbrook. N.J. (2004) Impact of land cover change on the climate of southwest Western Australia. Journal of Geophysical Research 109, D18109, doi:10.1029/2003JD004347.

Plantico, M.S., Karl, T.R., Kukla, G. and Gavin, J. (1990) Is recent climate change across the United States related to rising levels of anthropogenic greenhouse gases? Journal of Geophysical Research 95, D10, 16617–16637.

Pollack, H.N., Huang, S. and Shen, P.-Y. (1998) Climate change record in subsurface temperatures: A global perspective. Science 282, 279–281.

Pomerol, C. (Ed.) (1989) The Wines and Winelands of France: Geological Journeys. BRGM, Orléans.

Poore, R.Z., Quinn, T.M. and Verardo, S. (2004) Century-scale movement of the Atlantic Intertropical Convergence Zone linked to solar variability. Geophysical Research Letters 31, L12214, doi:10.1029/2004GL019940.

Portes, L. and Ruyssen, F. (1886) Traité de la Vigne et de Ses Produits, Tome 1. Octave Doin, Paris.

Prescott, J.A. (1965) The climatology of the vine – Vitis vinifera L. The cool limit of cultivation. Transactions of the Royal Society of South Australia 89, 5–23.

Prescott, J.A. (1969) The climatology of the vine (Vitis vinifera L.). 3. A comparison of France and Australia on the basis of the warmest month. Transactions of the Royal Society of South Australia 93, 7–15.

Price, S.F., Breen, P.J. Valladao, M. and Watson, B.T. (1995) Cluster sun exposure and quercetin in Pinot noir grapes and wine. American Journal of Enology and Viticulture 46, 187–194.

Price, S.F., Watson, B.T. and Valladao, M. (1996) Vineyard and winery effects on wine phenolics – flavonols in Oregon Pinot Noir. Proceedings, Ninth Australian Wine Industry Technical Conference, Adelaide 1995, 93–97.

Puillat, V. (1888) Mille Variétés de Vignes, third edition. Camille Coulet, Montpellier.

Ramanathan, V. (1988) The greenhouse theory of climate change: a test by an inadvertent global experiment. Science 240, 293–299.

Ramanathan, V., Callis, L. and 9 others (1987) Climate-chemical interactions and effects of changing atmospheric trace gases. Reviews of Geophysics 25, 1441–1482.

Ramanathan, V., Cess, R.D., Harrison, E.F., Minnis, P., Barkstrom, B.R., Ahmad, E. and Hartmann, D. (1989) Cloud-radiative forcing and climate: Results from the Earth Radiation Budget Experiment. Science 243, 57–63.

Rankine, B.C., Fornachon, J.C.M., Boehm, E.W. and Cellier, K.M. (1971) Influence of grape variety, climate and soil on grape composition and on the composition and quality of table wines. Vitis 10, 33–50.

Rapp, A. and Versini, G. (1991) Influence of nitrogen compounds in grapes on aroma compounds in wines. Proceedings, International Symposium on Nitrogen in Grapes and Wine, Seattle 1991, 156–164.

Rasool, S.I. and Schneider, S.H. (1971) Atmospheric carbon dioxide and aerosols: effects of large increases on global climate. Science 173, 138–141.

Ray, C. (1968) Lafite. Peter Davies, London.

Rayner, N.A., Brohan, P., Parker, D.E., Folland, C.K., Kennedy, J.J., Vanicek, M., Ansell, T.J. and Tett, S.F.B. (2006) Improved analyses of changes and uncertainties in sea surface temperature measured in situ since the mid-nineteenth century: the HadSST2 dataset. Journal of Climate 19, 446–469.

Rebucci, B., Poni, S., Intrieri, C., Magnanini, E. and Lakso, A.N. (1997) Effects of manipulated grape berry transpiration on post-veraison sugar accumulation. Australian Journal of Grape and Wine Research 3, 57–65.

Reddy, K.R., Hodges, H.F. and McKinion, J.M. (1997) Modeling temperature effects on cotton internode and leaf growth. Crop Science 37, 503–509.

Reeve, J.R., Carpenter-Boggs, L., Reganold, J.P., York, A.L., McGourty, G. and McCloskey, L.P. (2005) Soil and winegrape quality in biodynamically and organically managed vineyards. American Journal of Enology and Viticulture 56, 367–376.

Rendu, V. (1857) Ampélographie Francaise. Victor Masson, Paris.

Reynolds, A.G. and Wardle, D.A. (1997) Flavour development in the vineyard: impact of viticultural practices on grape monoterpenes and their relationship to wine sensory response. South African Journal for Enology and Viticulture 18(1), 3–18.

Ribéreau-Gayon, P. (Ed.) (1990) The Wines and Vineyards of France: A Complete Atlas and Guide. Viking Penguin, London.

Richards, D. (1983) The grape root system. Horticultural Reviews 5, 127–168.

Richardson, G.R. and Cowan, A.K. (1995) Abscisic acid content of Citrus flavedo in relation to colour development. Journal of Horticultural Science 70, 769–773.

Rind, D., Goldberg, R. and Ruedy, R. (1989) Change in climate variability in the 21st century. Climatic Change 14, 5–37.

Robertson, A., Overpeck, and 8 others (2001) Hypothesized climate forcing time series for the last 500 years. Journal of Geophysical Research 106(D14), 14783–14803.

Robinson, J. (1986) Vines, Grapes and Wines: the Wine Drinker's Guide to Grape Varieties. Mitchell Beazley, London.

Robinson, J. (2006) In 'The Oxford Companion to Wine', third edition, ed. J. Robinson, various entries on regions and grape varieties. Oxford University Press, Oxford.

Robinson, J.B. (1992) Grapevine nutrition. In 'Viticulture Volume 2 – Practices', first edition, ed. B.G. Coombe and P.R. Dry, 178–208. Winetitles, Adelaide.

Robinson, S. and Walker, M. (2006) When do grapes make tannins? Australian and New Zealand Grapegrower and Winemaker No. 509a (Annual Technical Issue), 97–105.

Robock, A. and Mao, J. (1992) Winter warming from large volcanic eruptions. Geophysical Research Letters 19, 2405–2408.

Roderick, M.L. and Farquar, G.D. (2002) The cause of decreased pan evaporation over the past 50 years. Science 298, 1410–1411.

Rogers, H.H., Peterson, C.M., McCrimmon, J.N. and Cure, J.D. (1992) Response of plant roots to elevated atmospheric carbon dioxide. Plant, Cell and Environment 15, 749–752.

Rogiers, S.Y., Keller, M., Holzapfel, B.P. and Virgona, J.M. (2000) Accumulation of potassium and calcium by ripening berries on field vines of Vitis vinifera (L.) cv. Shiraz. Australian Journal of Grape and Wine Research 6, 240–243.

Rogiers, S.Y., Smith, J.A., White, R., Keller, M., Holzapfel, B.P. and Virgona, J.M. (2001) Vascular function in berries of *Vitis vinifera* (L.) cv. Shiraz. Australian Journal of Grape and Wine Research 7, 46–51.

Rogiers, S.Y., Greer, D.H., Hatfield, J.M., Orchard, B.A. and Keller, M. (2006) Solute transport into Shiraz berries during development and late-ripening shrinkage. American Journal of Enology and Viticulture 57, 73–80.

Rossow, W.B. and Schiffer, R.A. (1999) Advances in understanding clouds from ISCCP. Bulletin of the American Meteorological Society 80, 2261–2287.

Röthlisberger, F. (1986) 10 000 Jahre Gletschergeschichte der Erde. Sauerländer, Aarau. (Cited by Wigley and Kelly 1990.)

Rowe, R.N. (1993) The growth of grapevines: the role of roots. Australian and New Zealand Wine Industry Journal 8, 326–328.

Ruhl, E.H. (1992) Effect of K supply and relative humidity on ion uptake and distribution on two grapevine rootstock varieties. Vitis 31, 23–33.

Saayman, D. and Kleynhans, P.H. (1978) The effect of soil type on wine quality. Proceedings, South African Society for Enology and Viticulture, 21st annual meeting, Cape Town 1978, 105–119.

Sacher, J.A. (1973) Senescence and postharvest physiology. Annual Review of Plant Physiology 24, 197–224.

SAWS (South African Weather Service) (2003) Climate data for the Western Cape. Private communication 27 March 2003.

Scafetta, N. and West, B.J. (2005) Estimated solar contribution to the global surface warming using the ACRIM TSI satellite composite. Geophysical Research Letters 32, L18713, doi:10.1029/2005GL023849.

Scafetta, N. and West, B.J. (2006) Phenomenological solar contribution to the 1900–2000 global surface warming. Geophysical Research Letters 33, L05708, doi:10.1029/2005GL025539.

Schlesinger, M.E. and Ramankutty, N. (1992) Implications for global warming of intercycle solar irradiance variations. Nature 360, 330–333.

Schneider, S.H. (1975) On the carbon dioxide-climate confusion. Journal of Atmospheric Sciences 32, 2060–2066.

Schove, D.J. (1955) The sunspot cycle, 649 B.C. to A.D. 2000. Journal of Geophysical Research 60, 127–146.

Schultz, H.R. (1993) Photosynthesis of sun and shade leaves of field grown grapevine (*Vitis vinifera* L.) in relation to leaf age. Suitability of the plastochron concept for expression of physiological age. Vitis, 32, 197–205.

Schultz, H.R. (2000) Climate change and viticulture: a European perspective on climatology, carbon dioxide and UV-B effects. Australian Journal of Grape and Wine Research 6, 2–12.

Schwartz, S.E. (1988) Are global cloud albedo and climate controlled by marine phytoplankton? Nature 336, 441–445.

Schwerdtfeger, W. (Ed.) (1976) World Survey of Climatology, Volume 12: Climates of Central and South America. Elsevier, Amsterdam.

Seguin, G. (1983) Influence des terroirs viticoles sur la constitution et la qualité des vendages. Bulletin de l'O.I.V. 56, 3–18.

Seguin, G. (1986) Terroirs and pedology of wine growing. Experentia 42, 861–873.

Senior, C.A. and Mitchell, J.F.B. (1993) Carbon dioxide and climate: the impact of cloud parameterization. Journal of Climate 6, 393–418.

Shen, P.Y., Pollack., H.N., Huang, S. and Wang, K. (1995) Effects of subsurface heterogeneity on the inference of climate change from borehole temperature data: Model studies and field examples from Canada. Journal of Geophysical Research 100 (B4), 6383–6396.

Shindell, D.T., Rind, D., Balachandran, N., Lean, J. and Lonergan, P. (1999) Solar cycle variability, ozone, and climate. Science 284, 305–308.

Shindell, D.T., Schmidt, G.A., Mann, M.E., Rind, D. and Waple, A. (2001) Solar forcing of regional climate during the Maunder Minimum. Science 294, 2149–2152.

Shindell, D.T., Schmidt, G.A., Miller, R.L. and Mann, M.E. (2003) Volcanic and solar forcing of climate change during the preindustrial era. Journal of Climate 16, 4094–4107.

Shine, K.P., Derwent, R.G., Wuebbles, D.J. and Morcrette, J.-J. (1990) Radiative forcing of climate. In 'Climate Change: the IPCC Scientific Assessment', ed. J.T. Houghton, G.J. Jenkins and J.J. Ephraums. Cambridge University Press, Cambridge.

Silverman, S.M. (1992) Secular variation of the aurora for the past 500 years. Reviews of Geophysics 30, 333–351.

Simon, A. (1971) The History of Champagne. Octopus, London.

Sinton, T.H., Ough, C.S., Kissler, J.J. and Kasimatis, A.N. (1978) Grape juice indicators for prediction of potential wine quality. 1. Relationship between crop level, juice and wine composition, and wine sensory ratings and scores. American Journal of Enology and Viticulture 29, 267–271.

Smart, R.E. (1992) Canopy management. In 'Viticulture Volume 2 – Practices', first edition, ed. B.G. Coombe and P.R. Dry, 85–103. Winetitles, Adelaide.

Smart, R.E. (2002) Yield limits for vineyards – who needs them? Australian and New Zealand Wine Industry Journal 17 (1), 28–30.

Smart, R.E. (2006) Benefits to industry stemming from microvinification at Tamar Ridge. Australian and New Zealand Wine Industry Journal 21 (5), 15–17.

Smart, R.E. and Robinson, M. (1991) Sunlight into Wine: a Handbook for Winegrape Canopy Management. Winetitles, Adelaide.

Smart, R.E. and Sinclair, T.R. (1976) Solar heating of grape berries and other spherical fruits. Agricultural Meteorology 17, 241–259.

Smith, T.M. and Reynolds, R.W. (2003) Extended reconstruction of global sea surface temperatures based on COADS data (1854–1997). Journal of Climate 16, 1495–1510.

Smithsonian Institute (1951) The Smithsonian Meteorological Tables, sixth edition. Washington.

Soar, C.J., Sadras, V.O. and Petrie, P.R. (2008) Climate drivers of red wine quality in four contrasting Australian wine regions. Australian Journal of Grape and Wine Research 14, 78–90.

Solanki, S.K. and Fligge, M. (1999) A reconstruction of total solar irradiance since 1700. Geophysical Research Letters 26, 2465–2468.

Solanki, S.K. and Krivova, A. (2003) Can solar variability explain global warming since 1970? Journal of Geophysical Research 108, A5, 1200, doi:10.1029/2002JA009753.

Solanki, S.K., Schüssler, M. and Fligge, M. (2000) Evolution of the sun's large-scale magnetic field since the Maunder Minimum. Nature 408, 445–447.

Solanki, S.K., Usoskin, I.G., Kromer, B., Schüssler, M. and Beer, J. (2004) Unusual activity of the sun during recent decades compared to the previous 11,000 years. Nature 431, 1084–1087.

Somers, T.C. (1971) The polymeric nature of wine pigments. Phytochemistry 10, 2175–2186.

Somers, T.C. (1977) A connection between potassium levels in the harvest and relative quality in Australian red wines. Australian Wine, Brewing and Spirit Review, 24 May 1977, 32–34.

Somers, T.C. (1998) The Wine Spectrum. Winetitles, Adelaide.

Somerville, R.C.J. and Remer, L.A. (1984) Cloud optical thickness feedbacks in the CO_2 climate problem. Journal of Geophysical Research 89 (D6), 9668–9672.

Spayd, S.E., Tarara, J.M., Mee, D.L. and Ferguson, J.C. (2002) Separation of sunlight and temperature effects on the composition of Vitis vinifera cv. Merlot berries. American Journal of Enology and Viticulture 53, 171–182.

Sponholz, W.R. (1991) Nitrogen compounds in grapes, must and wine. Proceedings, International Symposium on Nitrogen in Grapes and Wine, Seattle 1991, 67–77.

Srinivasan, C. and Mullins, M.G. (1980) Effects of temperature and growth regulators on formation of anlagen, tendrils and inflorescences in Vitis vinifera L. Annals of Botany 45, 439–446.

Srinivasan, C. and Mullins, M.G. (1981) Physiology of flowering in the grapevine – a review. American Journal of Enology and Viticulture 32, 47–63.

Staudt, M. and Bertin, N. (1998) Light and temperature dependence of the emission of cyclic and acyclic monoterpenes from holm oak (Quercus ilex L.) leaves. Plant, Cell and Environment 21, 385–395.

Stine, S. (1994) Extreme and persistent drought in California and Patagonia during mediaeval time. Nature 369, 546–549.

Stoll, M., Loveys, B. and Dry, P. (2000) Hormonal changes induced by partial rootzone drying of irrigated grapevine. Journal of Experimental Botany 51 (350), 1627–1634.

Stott, P.A., Tett, S.F.B., Jones, G.S., Allen, M.R., Mitchell, J.F.B. and Jenkins, G.J. (2000) External control of 20th century temperature by natural and anthropogenic forcings. Science 290, 2133–2137.

Stott, P.A., Jones, G.S. and Mitchell, J.F.B. (2003) Do models underestimate the solar contribution to recent climate change? Journal of Climate 16, 4079–4093.

Strauss, C.R., Wilson, B., Anderson, R. and Williams, P.J. (1987) Development of precursors of C_{13} nor-isoprenoid flavorants in Riesling grapes. American Journal of Enology and Viticulture 38, 23–27.

Stuiver, M., and Quay, P.D. (1980) Changes in atmospheric carbon-14 attributed to a variable sun. Science 207, 11–19.

Sturman, A. and Tapper, N. (1996) The Weather and Climate of Australia and New Zealand. Oxford University Press, Melbourne.

Sun, B. and Groisman, P.Ya. (2004) Variations in low cloud cover over the United States during the second half of the twentieth century. Journal of Climate 17, 1883–1888.

Svensmark, H. (1998) Influence of cosmic rays on Earth's climate. Physical Review Letters 81, 5027–5030.

Svensmark, H. and Friis-Christensen, E. (1997) Variation of cosmic ray flux and global cloud average – a missing link in solar-climate relationships. Journal of Atmospheric and Solar-Terrestrial Physics 59, 1225–1232.

Svensmark, H., Pedersen, J.O.P., Marsh, N.D., Enghoff, M.B. and Uggerhøj, U.I. (2007) Experimental evidence for the role of ions in particle nucleation under atmospheric conditions. Proceedings of the Royal Society A, 463, 385–396.

Swiegers, J.H., Bartowsky, E.J., Henschke, P.A. and Pretorius, I.S. (2005) Yeast and bacterial modulation of wine aroma and flavour. Australian Journal of Grape and Wine Research 11, 139–173.

Swiegers, J.H., Francis, I.L., Herderich, M.J. and Pretorius, I.S. (2006) Meeting consumer expectations through management in vineyard and winery: the choice of yeast for fermentation offers great potential to adjust the aroma of Sauvignon Blanc wine. Australian and New Zealand Wine Industry Journal 21 (1), 34–42.

Swinchatt, J. and Howell, D.G. (2004) The Winemaker's Dance: Exploring Terroir in the Napa Valley. University of California Press, Berkeley.

Symons, G.M., Davies, C., Shavrukov, Y., Dry, I.B., Reid, J.B. and Thomas, M.R. (2006) Grapes on steroids. Brassinosteroids are involved in grape berry ripening. Plant Physiology 140, 150–158.

Tanner, W. and Beevers, H. (1990) Does transpiration have an essential function in long-distance ion transport in plants? Plant, Cell and Environment 13, 745–750.

Tesic, D., Woolley, D.J., Hewett, E.W. and Martin, D.J. (2001a) Environmental effects on cv. Cabernet Sauvignon (Vitis vinifera L.) grown in Hawke's Bay, New Zealand. 1. Phenology and characterisation of viticultural environments. Australian Journal of Grape and Wine Research 8, 15–26.

Tesic, D., Woolley, D.J., Hewett, E.W. and Martin, D.J. (2001b) Environmental effects on cv. Cabernet Sauvignon (Vitis Vinifera L.) grown in Hawke's Bay, New Zealand. 2. Development of a site index. Australian Journal of Grape and Wine Research 8, 27–35.

Tett, S.F.B., Stott, P.A., Allen, M.R., Ingram, W.J. and Mitchell, J.F.B. (1999) Causes of twentieth-century temperature change near the Earth's surface. Nature 399, 569–572.

Tett, S.F.B., Jones, G.S. and 11 others (2002) Estimation of natural and anthropogenic contributions to twentieth century temperature change. Journal of Geophysical Research 107, D16, 10.1029/2000JD000028.

Thompson, L.G., Moseley-Thompson, E., Dansgaard, W. and Grootes, P.M. (1986) The Little Ice Age as recorded in the stratigraphy of the tropical Quelccaya Ice Cap. Science 234, 361–364.

Thomson, L.J. and Hoffman, A.A. (2008) Vegetation increases abundance of natural enemies of common pests in vineyards. Australian and New Zealand Grapegrower and Winemaker No. 533a (Annual Technical Issue), 34–37.

Timbal, B. and Arblaster, J.M. (2006) Land cover change as an additional forcing to explain the rainfall decline in the south west of Australia. Geophysical Research Letters 33, L07717, doi:10.1029/2005GL025361.

Tingey, D.T., Manning, M., Grothaus, L.C. and Burns, W.F. (1980) Influence of light and temperature on monoterpene emission rates from slash pine. Plant Physiology 65, 797–801.

Tingey, D.T., Turner, D.P. and Weber, J.A. (1991) Factors controlling the emissions of monoterpenes and other volatile organic compounds. In 'Trace Gas Emissions by Plants', ed. T.D. Sharkey, E.A. Holland and H.A. Mooney, 93–119. Academic Press, San Diego.

Tominaga, T., Baltenweck-Guyot, R., Peyrot des Gachons, C. and Dubourdieu, D. (2000) Contribution of volatile thiols to the aromas of white wines made from several *Vitis vinifera* grape varieties. American Journal of Enology and Viticulture 51, 178–181.

Torok, S.J., Morris, C.J.G., Skinner, C. and Plummer, N. (2001) Urban heat island features of southeast Australian towns. Australian Meteorological Magazine 50, 1–13.

Treeby, M., Holzapfel, B.P. and Friedrich, C.J. (1996) Managing vine nitrogen supply to improve wine grape composition. Proceedings, seminar 'Quality Management in Viticulture', Mildura 1996, 14–16. Australian Society of Viticulture and Oenology, Adelaide.

Trought, M. (2005) Fruitset – possible implications on wine quality. Proceedings, seminar 'Transforming Flowers to Fruit', Mildura 2005, 27–31. Australian Society of Viticulture and Oenology, Adelaide.

Turnbull, M.H., Murthy, R. and Griffin, K.L. (2002) The relative impacts of daytime and night-time warming on photosynthetic capacity in *Populus deltoides*. Plant, Cell and Environment 25, 1729–1737.

Turnbull, M.H., Tissue, D.T., Murthy, R., Wang, X., Sparrow, A.D. and Griffin, K.L. (2004) Nocturnal warming increases photosynthesis at elevated CO_2 partial pressure in *Populus deltoides*. New Phytologist 161, 819–826.

Turner, P. and Creasy, G.L. (2003) Terroir: competing definitions and applications. Australian and New Zealand Wine Industry Journal 18 (6), 48–55.

Unwin, T. (1990) Saxon and early Norman viticulture in England. Journal of Wine Research 1 (1), 61–75.

Usoskin, I.G., Marsh, N., Kovaltsov, G.A., Mursula, K. and Gladysheva, O.G. (2004) Latitudinal dependence of low cloud amount on cosmic ray induced ionization. Geophysical Research Letters 31, L16109, doi:10.1029/2004GL019507.

Van Eimern, J. (1968) Soil climate relationships. The dependence of soil temperature on radiation, albedo, moisture and agricultural practices. Proceedings, World Meteorological Organization seminar, Melbourne, 491–520. Bureau of Meteorology, Australia, Melbourne.

Van Leeuwen, C., Friant, P., Choné, X., Tregoat, O., Koundouras, S. and Dubourdieu, D. (2004) Influence of climate, soil and cultivar on terroir. American Journal of Enology and Viticulture 55, 207–217.

Van Zyl, J.L. and van Huyssteen, L. (1984) Soil and water management for optimum grape yield and quality under conditions of limited or no irrigation. Proceedings, Fifth Australian Wine Industry Technical Conference, Perth 1983, 25–66.

Verdecchia, M., Visconti, G., Giorgi, F. and Marinucci, M.R. (1994) Diurnal temperature range for a doubled carbon dioxide concentration experiment: analysis of possible physical mechanisms. Geophysical Research Letters 21, 1527–1530.

Viala, P. and Vermorel, V. (1901–1904) Traité Général de Viticulture: Ampélographie. Masson et Cie, Paris.

Von Storch, H., Zorita, E., Jones, J.M., Dimitriev, Y., González-Rouco, F. and Tett, S.F.B. (2004) Reconstructing past climate from noisy data. Science 306, 679–682.

Vose, R.S., Easterling D.R. and Gleason, B. (2005) Maximum and minimum temperature trends for the globe: An update through 2004. Geophysical Research Letters 32, L.23822, doi:10.1029/2005GL024379.

Wagner, G., Livingstone, D.M., Masarik, J., Muscheler, R. and Beer, J. (2001) Some results relevant to the discussion of a possible link between cosmic rays and the Earth's climate. Journal of Geophysical Research 106, D4, 3381–3387.

Wahl, K. (1988) Climate and soil effects on grapevine and wine: The situation on the northern border of viticulture – the example of Franconia. Proceedings, Second International Cool Climate and Oenology Symposium, Auckland 1988, 1–5.

Waldin, M. (2006a) Organic viticulture. In 'The Oxford Companion to Wine', third edition, ed. J. Robinson. Oxford University Press.

Waldin, M. (2006b) Biodynamic viticulture. In 'The Oxford Companion to Wine', third edition, ed. J.Robinson. Oxford University Press.

Weaver, R.J. (1962) The effect of benzo-thiazole-2-oxyacetic acid on maturation of seeded varieties of grapes. American Journal of Enology and Viticulture 13, 141–149.

Weckert, M. (2002) Vineyard microbial soil health. Australian and New Zealand Grapegrower and Winemaker No. 464, 21–24.

Weinhold, R. (1978) Vivat Bacchus: a History of the Vine and its Wine. Trans. from the German by Neil Jones. Argus, Watford.

Went, F.W. (1953) The effect of temperature on plant growth. Annual Review of Plant Physiology 4, 347–362.

Went, F.W. (1957) Experimental Control of Plant Growth. Chronica Botanica, Waltham, Mass.

Went, F.W. and Sheps, L.O. (1969) Environmental factors in regulation of growth and development: ecological factors. In 'Plant Physiology, a Treatise', ed. F.C. Steward, Vol. VA, 299–406. Academic Press, New York.

Wermelinger, B. (1991) Nitrogen dynamics in grapevine: physiology and modeling. Proceedings, International Symposium on Nitrogen in Grapes and Wine, Seattle 1991, 23–31.

White, R.E. (2003) Soils for Fine Wines. Oxford University Press, New York.

Whiting, J.R. (2004) Grapevine rootstocks. In 'Viticulture Volume 1 – Resources', second edition, ed. P.R. Dry and B.G. Coombe, 167–188. Winetitles, Adelaide.

Wigley, T.M.L. (1989) Possible climate change due to SO_2-derived cloud condensation nuclei. Nature 339, 365–367.

Wigley, T.M.L. and Kelly, P.M. (1990) Holocene climatic change, ^{14}C wiggles and variations in solar irradiance. Philosophical Transactions of the Royal Society of London A, 330, 547–560.

Wigley, T.M.L. and Raper, S.C.B. (1990) Climatic change due to solar irradiance changes. Geophysical Research Letters 17, 2169–2172.

Wiles, G.C., D'Arrigo, R.D., Villalba, R., Calkin, P.E. and Barclay, D.J. (2004) Century-scale solar variability and Alaskan temperature change over the past millennium. Geophysical Research Letters 31, L15203, doi:10.1029/2004GL020050.

Wilson, B., Strauss, C.R. and Williams, P.J. (1984) Changes in free and glycosidically bound monoterpenes in developing Muscat grapes. Journal of Agricultural and Food Chemistry 32, 919–924.

Wilson, J.E. (1998) Terroir: The Role of Geology, Climate and Culture in the Making of French Wines. Mitchell Beasley, London.

Winkler, A.J. (1962) General Viticulture. University of California Press, Berkeley. Australian edition 1963, Jacaranda Press, Brisbane.

Winkler, A.J., Cook, J.A., Kliewer, W.M. and Lider, L.A. (1974) General Viticulture, second edition. University of California Press, Berkeley.

Wittwer, S.H. (1990) Implications of the greenhouse effect on crop productivity. HortScience 25, 1560–1567.

WMO (World Meteorological Organization) (1998) 1961–1990 Global Climate Normals, CD-ROM. NCDC, USA, Asheville NC.

Wolf, T.K., Dry, P.R., Iland, P.G., Botting, D., Dick, J., Kennedy, U. and Ristic, R. (2003) Response of Shiraz grapevines to five different training systems in the Barossa Valley, Australia. Australian Journal of Grape and Wine Research 9, 82–95.

Wong, S.C. (1980) Effects of elevated partial pressure of CO_2 on rate of CO_2 assimilation and water use efficiency in plants. In 'Carbon Dioxide and Climate: Australian Research', ed. G.I. Pearman, 159–166. Australian Academy of Science, Canberra.

Wood, F.B. (1988) Comment: on the need for validation of the Jones et al. temperature trends with respect to urban warming. Climatic Change 12, 297–312.

Woodhouse, C.A. and Overpeck, J.T. (1998) 2000 years of drought variability in the central United States. Bulletin of the American Meteorological Society 79, 2693–2714.

WRCC (Western Region Climate Center, USA) (2005) Monthly climate summaries to 31 December 2004. By internet http://www.wrcc.dri.edu.

Yang, B., Braeuning, A., Johnson, K.R. and Yafeng, S. (2002) General characteristics of temperature variation in China during the last two millennia. Geophysical Research Letters 29 No. 9, 10.1029/2001GL014485.

Yu, F. (2002) Altitude variations of cosmic ray induced production of aerosols: implications for global cloudiness and climate. Journal of Geophysical Research 107, A7, 10.1029/2001JA000248.

Zheng, X., Basher, R.E. and Thompson, C.S. (1997) Trend detection in regional-mean temperature series: maximum, minimum, mean, diurnal range, and SST. Journal of Climate 10, 317–326.

Author Index

Index

Wakefield Press is an independent publishing and
distribution company based in Adelaide, South Australia.
We love good stories and publish beautiful books.
To see our full range of titles, please visit our website at
www.wakefieldpress.com.au.